T0309173

IMPLEMENTING STANDARDIZED WORK

PROCESS IMPROVEMENT

The One-Day Expert Series

Series Editor
Alain Patchong

The One-Day Expert Series

IMPLEMENTING STANDARDIZED WORK

PROCESS IMPROVEMENT

Alain Patchong

CRC Press
Taylor & Francis Group
Boca Raton London New York

CRC Press is an imprint of the
Taylor & Francis Group, an **informa** business

A PRODUCTIVITY PRESS BOOK

CRC Press
Taylor & Francis Group
6000 Broken Sound Parkway NW, Suite 300
Boca Raton, FL 33487-2742

Library of Congress Cataloging-in-Publication Data

Patchong, Alain.
  Implementing standardized work : process improvement / Alain Patchong.
      pages cm. -- (The one-day expert series)
  Includes bibliographical references and index.
  ISBN 978-1-4665-6358-2
  1. Production management. 2. Workflow. 3. Process control. I. Title.

TS155.P353 2014
658.5--dc23   2013040712

Visit the Taylor & Francis Web site at
http://www.taylorandfrancis.com

and the CRC Press Web site at
http://www.crcpress.com

Contents

Preface

THE ONE-DAY EXPERT

The One-Day Expert series presents subjects in the simplest way while maintaining the substance of the matter. This series allows anyone to acquire quick expertise in a subject in less than a day. That means reading the book, understanding the practical description given in the book, and applying it right away, in only one day. To focus on the quintessential knowledge, each *The One-Day Expert* book addresses only one topic and presents it through a streamlined, simple, narrative story. Clear and simple examples are used throughout each book to ease understanding and, thereafter, application of the subject.

Acknowledgments

The One-Day Expert series is the direct consequence of my previous work at Goodyear. I owe thanks to several of my former colleagues, who provided me with valuable remarks and comments.

I am thankful to Dariusz Przybyslawski and Mike Kipe, who were part of the team I formed to deploy Standardized Work in Goodyear plants. Dariusz's help was instrumental in structuring and tuning *The One-Day Expert* series.

I am very grateful to two other former colleagues at Goodyear who, at a very early stage, believed in Standardized Work as presented in this book and gave me the opportunity to try it on the shop floor: François Delé and Markus Wachter.

I am obliged to Alain Prioul, Xavier Oliveira, Pierre-Antoine Rappenne, and Philip Robinson, who have read drafts and offered valuable suggestions for improvement. I am also very thankful to the editorial staff of Taylor & Francis for their wonderful work improving the readability of the initial text.

I express my gratitude to all my colleagues at Faurecia who work with me to test ideas and actions and critique or support my thoughts.

Finally, and most especially, I would like to give my special thanks to my wife, Patricia; my son, Elykia; and my daughter, Anya, for their unrelenting support and patience.

1

Introduction

This book is the third on Standardized Work deployment and is dedicated to methods and tools that help to improve the process.* The previous two books dealt with the initial steps of Standardized Work deployment, which consist of capturing the current state. The first book addressed operator performance assessment; the second one was dedicated to tools that help capture the current state of the process itself: process analysis chart, standardized work combination table, standardized work chart, and operator work instructions.

As stated in the first book of this series, *Implementing Standardized Work (Operators' Performance Measurement)*, for standards to really be sustainable and beneficial, they have to be implemented as part of a systematic approach that includes many other activities. One of those activities is process improvement. In effect, Standardized Work documents alone make little sense if they are not used to foster progress through continuous improvement. Conversely, any improvement that is not consigned in standards might become elusive, difficult to convey, and daunting to assess. To summarize in a few words: Process improvement needs standardization, and standardization needs process improvement. In addition and most important, there is no better motivator than success. Process improvement results produce the quick wins† that are needed to keep momentum in the team and create opportunities to celebrate. As with any continuous improvement activity, implementing Standardized Work may be a long march. People will not go a long way if they are not convinced

---

* The first and the second books are, respectively, *Implementing Standardized Work: Operators' Performance Measurement* and *Implementing Standardized Work: Writing Standardized Work Forms*.
† This is what John Kotter calls short-term wins in his book *Leading Change* (Harvard Business School Press, Cambridge, MA, 1996).

that it is worthwhile. Without quick wins, most people will withdraw, and some may eventually become resistant. With the materialization of process improvement results, people start to see positive change, which is absolutely instrumental to keep the team motivated. Ultimately, this helps to cover the distance necessary to deliver the long-term benefits of Standardized Work: improved safety, better quality, higher productivity, reduced costs, and increased team morale—all of which are the foundation of operational excellence.

The main concern of the current book is how to generate the aforementioned quick wins, rather than more on Kaizen activities. On one hand, it shows how to use Standardized Work forms to identify easy and quick opportunities for improvement. On the other hand, it provides simple tools and forms to generate quick and easy-to-implement improvements to nurture the enthusiasm that sustains Standardized Work implementation. Process improvement for the implementation of Standardized Work is basically supported by two activities. The first is a sort of benchmark for similar activities handled by different workers. It consists of using time collection results, Standardized Work forms, pictures, and videos to initiate *Black Books Sharing*, which amounts to openly sharing operators' practices and identifying and gathering best ones. The second key activity is based on intensive observation called *Tachinbo*.* There are, in general, three main focal points to this intensive observation: equipment, method, and work conditions. Additional tools are proposed to support and channel these activities, prioritize outcomes to identify quick wins, and implement them straightaway.

In *The One-Day Expert Series* dedicated to Standardized Work, Thomas, a young, high-potential plant manager in an industrial group, is reassigned to another plant, which is losing money. Previous plant managers have tried several initiatives with, to say the least, limited results. His urgent mission, which sounds like the EMEA (Europe, Middle East, and Africa) senior management's last card, is simple: turn the plant around. The morale in the plant is low; the staff is equally pessimistic about the plant's future and distrustful of senior management. Time is running out; company headquarters needs concrete results and has become impatient. To face these challenges, Thomas has decided to use Standardized Work deployment to achieve quick and visible results while rebuilding a real team. To this end, he has requested the support of Daniel Smith,

---

* *Tachinbo* is a Japanese word that translates as "standing still" in English.

the industrial engineering manager for EMEA. Daniel had been with the company for only a couple of years, after previous experience in the automotive industry. Building on his previous experience, he recently designed and launched a Standardized Work initiative and is looking to prove the real power of Standardized Work by deploying it in several plants.

This series of books dedicated to Standardized Work improvement recounts, step by step, Thomas's deployment of Standardized Work with Daniel's support. The first book showed the initial steps Thomas took to assess the plant's current situation through measurement of operators' performance. The second book recounted the next steps of this assessment, which consisted of writing Standardized Work forms to help see both variability and waste. It also shed light on a dire industrial engineering community squashed by a ubiquitous Excellence System* organization and locked in an unrelentingly perilous fight for relevance, if not survival. In this third book, Thomas opens a new "front line" in his quest to turn around the plant as he tries a new type of relationship with the labor union based on mutual trust and constructive partnership while negotiating a competitiveness plan. We will also see how he continues to push for the implementation of Standardized Work. More precisely, we will learn about the use of Standardized Work documents to generate and implement simple and fast improvements that produce quick wins to keep the energy high while implementing Standardized Work.

---

* An Excellence System is a local Lean implementation organization.

2

There Is No Panacea!

Thomas has been living in the same hotel for weeks now. As a special customer, he arranged with the hotel manager to keep the same room. He had also been granted some privileges. For instance, as a very early coffee drinker, he got the right to access the kitchen any time he wanted to help himself. Over time, he had created a new home in his hotel, where, ironically, he was spending more time than at his own place. Thomas's peers considered him to be one of the smartest guys around. But, being smart was not enough; therefore, he worked as hard as possible to achieve his objectives. He never looked for any excuse for inaction or to explain his failures. Facing the choice between action with high risk of failure and inaction, Thomas would always go for action. He even had a catchphrase to motivate his staff when something went wrong: "The man who makes no mistakes does not usually make anything."[*] He thereby encouraged everyone to move on after a failure. As he often asserted: "Real failure is actually 'not trying' rather than 'not succeeding.'"

Back at his hotel room after three intensive days of Standardized Work training, Thomas had little sleep that night. While driving to his hotel, he had heard the U.S. president commenting on the most recent job creations, saying that the U.S. economy was starting to pick up steam. "The gears are starting to turn again, and we're getting some traction," the U.S. president said. Thomas thought, "No such thing could be said about France or Europe in general. With the jobless rate above 10% in France and almost three times higher in countries like Spain and Greece, Europe is clearly in a gloomy situation. Economic crisis is nothing new, but this time it looks different." The other day, he had a discussion with an old friend in charge of improvement in an automotive company's plants, who said something

---

[*] Quotation from Edward Phelps.

that was still fresh in his mind. "In the previous crises, people used to take advantage of the additional free time created by a drop in volume to get ready for better times. It was therefore not unusual to see managers launch multiple Kaizen events here and there to get prepared for the next cycle of growth. Things look very different these days. Most people do not see this as another crisis, but a 'new normal.' I can tell you that it has become very difficult to energize people through Kaizen projects."

Thomas had decided at a very young age to become an engineer because he adored tinkering and spending leisure time inventing little toys. Later, after graduating from one of his country's best schools, he spent his first years doing just what he loved: being an engineer working in manufacturing engineering, machine design, then production. He worked hard, had lot of fun doing his job, and would, from time to time, feel somewhat remorseful when interacting with humanitarian friends. As he often framed it, "Those folks spend their time changing people's lives for the better, while I am just content to have fun at my job." Things have changed drastically since his early professional life. The tableau of an abundant Western European society has morphed into a dire picture as industry is continuously contracting. To Thomas's eyes, the situation was pretty clear. Here, as elsewhere in Western countries, industry provides most of the decent middle-class jobs. Its contraction in response to economic strains was jeopardizing the whole social fabric of much of European society, if not the civilization itself. For all these reasons, Thomas considered himself more and more as a man on a mission, just like those humanitarians he had envied several years ago. He felt like his new assignment in this French plant gave him the opportunity to contribute his share, something more than a personal challenge.

Thomas had noticed that to sustain their margin and keep industry in Europe, a growing number of companies had been focusing efforts on improving their operations. As proof, he would say, "A term like 'Lean', which was not even known to people around me in the early 2000s, is now mentioned everywhere." To support his point, he would spontaneously refer to changes in a company's organization charts. "At the highest level, you now see a director or a vice president in charge of Lean stuff. Also, in each plant, you have someone or, more accurately, a team of people, in charge of continuous improvement. Europe has embraced Lean methods and is using them in every part of the business: Lean in manufacturing, in project management, in human resource management, in offices, in finance, in education, you name it. What huge progress from a few years ago!" Actually, Thomas had also been chilled by some dogmatic approaches, and he would talk endlessly about how harmful they could be in the long term. He would explain that even Toyota had been evolving some of its initial principles.

His best example would be what Daniel called the "zero-buffer" dogma. Taiichi Ohno, a Toyota engineer who is considered the father of the Toyota Production System or Lean manufacturing, reportedly considered buffers absolute evil. Therefore, buffers were completely forbidden. During his reign, Toyota would build very long lines with no buffers. As a result, the whole line would stop every time a workstation stopped. This was arguably a means to put pressure on people to solve the root cause of problems and not to hide them in inventories. However, overall, this would lead to poor line efficiency. After Ohno's retirement, Toyota folks decided to amend this principle and began including small buffers in their lines.

The discussion Thomas had with Daniel when he stopped by his office a few hours ago was still fresh in his mind. Daniel started by drawing a curve on a flip chart (chart 1 of Figure 2.1) and commented: "It is true that people's reactivity decreases as buffer size increases. Because people see lots of parts in the buffer, they tend to fall into some idleness and fail to see the urgency of fixing problems that occur on the line. This is something I have noticed myself in my daily work over several years of experience. It is also confirmed by manufacturing scientists.[*] Based on this metric, Taiichi Ohno was right to distain buffers." While pursuing the explanation, you could tell from the tone of his voice that Daniel was leading up to a "but." This came when he started drawing another chart (chart 2 of Figure 2.1):

> But, what Toyota learned the hard way was that the zero-buffer approach leads to a catastrophe in efficiency. An analytic model of a production line shows a curve of efficiency that looks like this one I just drew[†] [chart 2 of Figure 2.1]. It is easy to see that adding a small buffer to a zero-buffer line will drastically increase its efficiency, which means that there is a huge

---

[*] Stanley Baiman, Serguei Netessin, and Richard Saouma, Informativeness, Incentive Compensation, and the Choice of Inventory Buffer, *The Accounting Review* Vol. 85, No. 6, pp. 1839–1860 (2010); and Kenneth L. Schultz, David C. Juran, John W. Boudreau, John O. McClain, and L. Joseph Thomas, Modeling and Worker Motivation in JIT Production Systems, *Management Science* Vol. 44, No. 12, Part 1 of 2 (1998).

[†] For more details on analytic models, refer to Stanley B. Gershwin, *Manufacturing Systems Engineering*, PTR Prentice Hall, Englewood Cliffs, NJ, 1994.

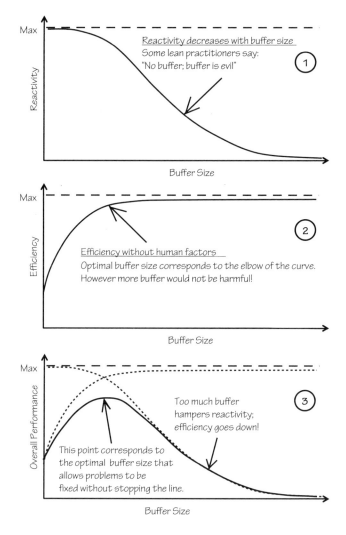

FIGURE 2.1
Factory physics, which combines human factors and a production line analytic model, shows that a small buffer has a huge impact on the efficiency of a line, while too much buffer has an overall negative effect.

incentive to add a small-size buffer. However, this curve does not tell the whole story, as it tends to show that more buffers are not harmful to the line efficiency even though the payoff is less.

Daniel moved on and depicted a third graph while explaining the following:

Actually, the correct picture is reached by combining the two previous graphs to get a more holistic approach [chart 3 of Figure 2.1]. That means including human factors into the analytic model to take into account the fact that a worker's reaction depends on the size of the buffer. This is because we are looking at the behavior of the whole system, including the workers and the line. It would be incomplete to focus only on people's reactions or to study the line only, without human factors. When we consider this worker-line system, as you can see, an optimal buffer size appears. This buffer is small enough to keep the reactivity high and big enough to allow workers to fix most of the problems that may occur without stopping the line. Very prosaically, a buffer should be looked at as a medicine that, like any medicine, has side effects. In consequence, when too much of it is used, the negative parts trump the positive parts. The irony of this situation is that even years after Toyota moved away from the "zero-buffer dogma," many Lean practitioners are still preaching about it zealously. The same folks will tell you that all you have been doing in your job is wrong and urge you to convert to Lean instantly, whatever your job: manufacturing, finance, management. ... A part of what they say is absolutely true, as some basic Lean principles can help you improve a lot. However, bear in mind that the Toyota Production System, or Lean, will not solve all your problems. You should think critically. Here's the thing: There is no panacea!

On the memory of those words, Thomas tried to fall asleep, convinced that if he was to turn around the plant, implanting Lean tools alone will not suffice. He needed something else, maybe a sort of give-and-take composite plan. He had a quick look at his clock; it was 2 a.m. He understood that the night would be very short. "Not the best way to get ready for another intensive day of workshop," he muttered.

Toyota Production System or Lean, will not solve all your problems. You should think critically. Here is the thing: There is no panacea!

3

Training Day

Thursday was the fourth day of training. As always, Daniel was first to the training room. He needed this time to prepare before the training, before the bulk of the participants arrived. The ritual was clearly defined. As always, he would need to spend some time discussing yesterday's training. He would keep asking himself: "Were the main messages conveyed and key learning assimilated?" Daniel would then say a few words about the day's training, just like a chef announcing the menu to his guests. "This allegory resounded well in this country well renowned for its cuisine," Daniel thought. This consisted of laying down the content of the day's training in his now-well-known, step-by-step structure (Figure 3.1), still with the tantalizing touch of a chef.

Unlike the previous day, Thursday was a very sunny day. In this area of France, it was not unusual to move from autumn to spring weather within a few hours. Unlike Wednesday, everyone had arrived before the beginning of the training. After a glance at the room, Daniel noticed that, despite the nice weather, some faces showed the toll taken from the last three intensive days. He therefore understood that today would be more challenging than the previous days. This was a common pattern he had noticed throughout all the training he had given so far. "Thursday is always the most difficult day, the make-or-break day, which can sink your whole week's training; you need to be careful," he murmured to himself. To balance the physical weariness of the bodies, Daniel's method consisted of appealing to their mental energy by raising the level of enthusiasm. Nothing could do better than real success achieved on the shop floor. This need was in perfect sync with the planning, as Thursday was dedicated to process improvements that would provide some quick wins in the end.

Not all faces looked tired. Among the brighter ones was Eric's. The young Excellence System manager was accustomed to weeklong intensive trainings.

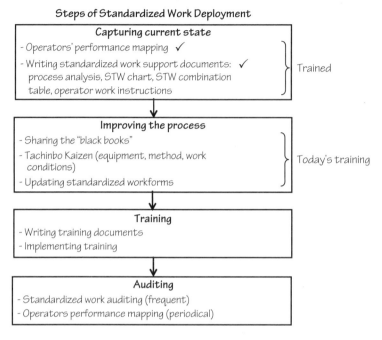

FIGURE 3.1
Steps of Standardized Work deployment.

Eric was no stranger to Daniel. Since he joined the company, they had met several times. As one of Daniel's colleagues once put it: "This fellow had enough energy to heat the whole city for a while." Eric, who was in his midthirties, joined the company 8 years ago, after a sparkling career in a small plastics company. At least this is how he would voluntarily describe it. Everyone around him had heard several stories, in which he liked to brag about being a plant manager in his midtwenties. Despite his tendency to endlessly boast about his past achievements, Eric was unanimously recognized as a leader who would not hesitate to push everybody around while getting his hands dirty to make things happen. Eric was also considered a hypercompetitive person; whatever he was doing, he played to win. He therefore took every occasion when he or his team would be compared to others seriously. This attitude also applied to late-evening team-building bowling parties, when everyone else simply wanted to have fun. Here as well, losing was not an option.

It was no surprise that Eric felt very uncomfortable being part of a team running one of the worst plants in the region. He felt like his plant's performance was nowhere near the kind of standards he set for himself. He expressed this discomfort on several occasions, for he generally spoke his mind. As a straight talker—some said straight shooter—he was perceived rather positively by those around him, although some complained that too much of it made him appear like a schoolyard bully. Talent spotted by the EMEA (Europe, Middle East, and Africa) senior management, Eric was rumored to be moving to headquarters after his current assignment. In the meantime, he had a perilous task: contribute to the turnaround of his current plant. "Not a piece of cake," as he would most certainly say.

In this plant, as in many others, the friction, to say the least, between the industrial engineering (IE) and nascent Excellence System teams had left some unhealed wounds. Relations between the very analytic Steve, the IE manager, and the energetic straight talker Eric were not at their best, not quite at the level of hostility between two heavyweight boxers, but there was a perceptible frost. However, they were two smart guys who could arrange to work together even in some tense situations. The dire situation of the plant left no chance for distraction; the Standardized Work system, as designed by Daniel, would need all of them to work as a team. Therefore, to be successful, they needed to put aside their personal uneasiness with each other. Bringing IE and Excellence System people together to work for a common goal had been one of the least-anticipated challenges Daniel faced wherever the training was delivered. Daniel knew that he could count on Eric for anything, especially on this crucial day dedicated to process improvement.

As expected, the session started with the review of the previous day's learning. "What did you retain from yesterday's training?" Daniel asked. Steve jumped in:

> A big chunk of our core business is about writing standards. I must say that I was surprised at how easy it appeared: your step-by-step approach and your advice to focus not on the time, but on the method used by workers. As you said, "The right method would lead to the right time." I have to admit that we IE people tend to focus a lot on the time. I appreciated that you did not request that we get rid of our current standards because, as you underscored correctly, we still need them. When it comes to the charts, I agree with you that they are way more visual, and the warm-up exercise was very convincing. The *Yamazumihyo*˙ was new to me. It looked like a very powerful chart for process analysis.

˙ Yamazumihyo is actually the concentration of three words in Japanese: Yama, which means mountain; Zumi, which means accumulate; and Hyo, which means chart. Like many Japanese companies, we use the word "Yamazumihyo" to characterize the Process Analysis chart. However, it may be used to designate something else in some other Japanese companies. For more details, please refer the second book of the series: Implementing Standardized Work: Writing Standardized Work Forms.

Steve, who felt like he was talking too much, paused and concluded, "That's it, as far as I am concerned. I will let the other participants complete." While looking toward Eric, the Excellence System manager, Daniel continued: "Please go ahead and let us know your takeaway from yesterday." Eric had participated in another training on Standardized Work forms writing. He was well positioned to compare what he had learned before to what Daniel had taught yesterday.

> I very much liked your step-by-step approach. I found what you said about Key Points to be very instructive. I have to say that when I first got the training about how to find them, it looked very fuzzy to me. For sure, you did not give us an equation to find them, but I do have a better understanding now. Also, I understand their importance and why you decided to make them one of the elements of Standardized Work. Just like Steve, I like the Yama (short for *Yamazumihyo*) chart as well. It was new stuff for me too. I am very interested to see how we will be using our charts today to generate improvement ideas.

After Eric's words, it was Sarah's turn to give her feedback. Sarah, the human resources (HR) manager who was also in charge of training, insisted that focusing on the method instead of the result was the right thing. As she commented: "Well, it will change the way we address problems from 'finger pointing to problem solving.' I consider this a big shift in our mindset. It also puts more responsibility on the training, which I am in charge of. My understanding so far is that we need to focus on finding the right method and train our people to master it. I am very okay with that. This is what I have been preaching for a while, and I am glad it appeared in yesterday's key messages." John, the engineering manager, also shared a few words: "All this is new to me. However, I found it very easy to understand. I will be looking forward to seeing the kind of improvements that happen a lot today and to seeing how we can contribute to the implementation."

At this point, lots had been said. Thomas felt no need to repeat the points. He then raised a rather general point about what he called Daniel's philosophy, which included taking the liberty to modify and customize tools seen elsewhere to best fit the situation at hand. He concluded by saying: "This is the best way to learn as well. I am not sure that we are really making progress when we put ourselves in the mindset of believers and consider everything that comes out of Toyota's playbook as a dogma to be applied without question."

David, the person in charge of HSE (health, safety, and environment) and ergonomics at the plant, had something to share as a follow-up to an informal discussion he had with Daniel two days before. The point was about the impact of Standardized Work on safety. Safety was the company's biggest focus, whose slogan was "Safety first!" During this exchange, David explained that the company had invested a lot of effort into improving safety, with visible progress. Daniel agreed, but added that: "The safety improvement indicator has been somewhat flat recently." He shared a suggestion that he had been pushing at the regional level. "I do believe Standardized Work implementation and further safety progress are inextricably interwoven. I think it is the new frontier in our fight to improve workers' safety." To make his case, he asked if David could come up with a chart of the causes of accidents in his plant over the last 5 years. After two long nights working on the subject, David had something to share with the audience (Figure 3.2). The chart showed that most (67%) accidents that happened over the last few years were due to unsafe acts. "Identifying the safest way to do the job, including them in the key points, and training our people could really make the difference here," David asserted convincingly. Daniel nodded and commented: "The second cause, namely,

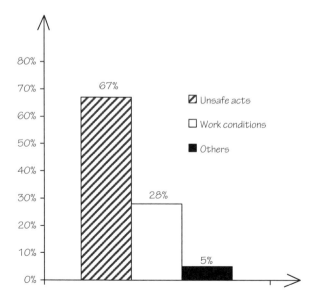

FIGURE 3.2

Most (67%) of accidents in a plant are due to unsafe acts. Implementing Standardized Work helps reduce worker injuries by decreasing unsafe acts and improving work conditions.

'work conditions,' can be improved through the implementation of Standardized Work as well. I will come back to this point later. So, overall, it is pretty clear that the implementation of Standardized Work will have a big impact on safety in our plants." Everybody in the room seemed to concur.

Visibly pleased by the start of the day, Daniel thanked the participants for their feedback and congratulated David for the chart he had shared. He noticed that, compared to the previous day, more things had been said in this morning's recap. Was that a sign of increased involvement from the trainees, which meant better understanding? He could not know for sure. Rather than trying to answer, he chose to wait and see. "Let's see how it unfolds today," he whispered to himself. He also decided that he would underscore once more why it was necessary to deploy Standardized Work as a system composed of four steps: capturing the current state, improving the process, training, and auditing. He explained that each of these four steps, deployed individually, would yield few results that could fade very quickly, whereas a system approach would provide the fabric that would strengthen the deployment and amplify the magnitude of the results.

It was time to move to the day's agenda. In a now-ritual gesture, Daniel moved to the chart he presented on day one (Figure 3.1) and commented:

We have now covered the first block of this chart, which consists of capturing the current state. Today's training will focus on improving the process. I would like to elaborate a little bit more on the "improving-the-process" block shown rather succinctly on this day one chart [Figure 3.1]. This is actually a multistep process. The first step will be about using data collected on the shop floor and Standardized Work forms to support best practice sharing that we also call *black books* sharing. I need to say that video recording can be helpful as well, especially for beginners. Some of you who were not here the first day might be asking yourself what *black books* means. Well, I will come back to this later. The second step consists of an intensive observation called *Tachinbo*. Yes, another Japanese word! This intensive observation, along with the use of previously written Standardized Work forms, will lead to a list of improvement actions regarding work conditions, equipment, and work method. These actions will be prioritized based on foreseen benefits and efforts. In this process, actions that require extensive effort will be eliminated, and priority will be given the ones that require the least effort. The result of this step is an action plan. The easiest actions to implement will be executed today or within a few days. These actions are expected to yield quick results with little manpower or investment effort. Before we proceed, the time is right to mention some caveats. Some of you have already been through a Kaizen workshop. This is not a Kaizen workshop *per se*. If we were to do such a workshop, we would need another week for that. Process improvement activities are included here as part of the Standardized Work system approach, as I explained previously. Our main objective today is to grasp some just-do-it actions and other quick ones, which I called zero-CAPEX* actions the first day while introducing the Standardized Work deployment. These actions are necessary to keep the motivation high in the team, as they give a sense of achievement after some hard work. Without them, writing Standardized Work forms may sound senseless. The final step of our work will then be to update Standardized Work documents to close the loop. Clearly, you need a standardized process to see opportunities for improvement. As Taiichi Ohno famously said, let me paraphrase him again: "There is no improve-ment without standardization." This is what we did yesterday by writing Standardized Work forms. On the other hand, you need standardization to harvest or realize your improvement ideas. This is what we will be doing at the end of today. To sum up, we do need standardization (for instance, standardized work forms) before and after improvement. Finally, as you can see on the chart I printed out for you [Figure 3.3], standardization and improvement are alternating processes, which are flip sides of the same coin. They ultimately lead to improved safety, better quality and productiv-ity, and therefore more dollars in your pocket.

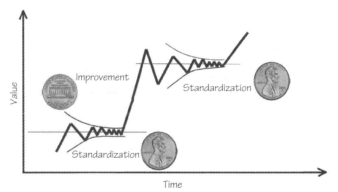

Standardization and improvement are flip sides of the same coin.
They ultimately lead to improved safety, better quality, and higher productivity,
hence more dollars in your pocket over time.

FIGURE 3.3
Standardization and improvement are the two sides of the same coin.

---

* These are actions whose implementation requires very little capital expenditure (CAPEX).

Daniel stopped for a few seconds and stared at the room to make sure that everything was all right, then proceeded. "One last point before I conclude. Although it sounds obvious, it is worth reminding that, in this third part, we will also be looking at opportunities to extend any improvement made to any other similar processes in the plant or elsewhere to rack up the maximum benefit. Well, this is all I have to tell you about today's agenda. As always, if there is anything unclear in your mind, please stop and ask! Now guys, let's get started. Thomas, would you like to summarize as usual?" Thomas jumped to the paperboard and quickly sketched a chart to illustrate his summary (Figure 3.4). He explained: "This is how I see the process when it comes to the four operators of our four teams*: With black books sharing, we will go

FIGURE 3.4
Steps of process improvement in Standardized Work deployment.

---

* This refers to the four teams formed for the simulation exercise, which is used to exemplify each step of training. For more details, please refer to the previous books in this series.

from four different work methods to only one, which I call 'Worker Standard.' This is the combination of best practices from four operators. Then, Tachinbo and prioritization, execution, and updating standardized forms will follow." Daniel thanked Thomas and added: "I like your chart. It tells the story exactly." He then stopped, looked around, and continued: "Now, it is time to delve deep into each of those tools. We will start with the 'black book sharing.' As always, explanation will be followed by application on the T-shirt simulation we have been using from the beginning."

There is no improvement without standardization. Standardization without improvement is senseless. Therefore, standardization and improvement are the flip sides of the same coin.

4

Sharing Black Books

Daniel thought that he had to contextualize the phrase "black book." "Does anybody here know the meaning of a black book?" Two or three people raised their hands. Daniel asked one of them to explain. "A *black book* is a book of names of people who should be censored or punished for some reason," the person answered. "That is absolutely correct; actually, it is one of the meanings of black book, but not really the one we will be using," Daniel responded. "There are other meanings when you look in a dictionary, but the one that I will be introducing here is that of Brian Joiner.* We'll call a black book a set of practices or tips that each worker has gathered through the years to ease or more broadly improve his or her work. Most of the time, those tips are a jealously kept secret by the owner and never shared with other people." Daniel paused for a few seconds to underscore the next sentence. "The word *black* here is related to the secrecy of what is contained in the book. I am sure that you have seen several examples in this plant or elsewhere. Could you share with us?"

John, the engineering manager, had a story to share. "I used to work in the body shop of a carmaker," he started.

> As some of you may know, the main operation in a body shop is welding steel parts. In most of the plants, this operation is performed by robots, which are controlled by programs. A few days after I joined this company, I first noticed some strange operations at the beginning of the afternoon shift. Every day, before starting his shift, a group leader of an afternoon team would move from PLC [programmable logic controller] to PLC with a floppy disk that he would introduce in the reader and make some weird

---

* Brian L. Joiner, *Fourth Generation Management: The New Business Consciousness*, McGraw-Hill, New York, 1994.

manipulations. Puzzled by such activities, I decided to check what was going on in other areas before discussing with my boss. I noticed the same kind of activities in other areas and later in almost all shifts. I therefore decided to discuss this peculiar behavior with my boss. Actually, when I recounted the story to him, he almost laughed in my face. He explained to me that each team had its own robot programs that suited their own way of production and would not trust those of other teams. Therefore, at the beginning of each team's shift, the team would erase the previous shift's robot programs and upload their own programs developed by their own robotics technicians, which included their secret findings and good practices.

Daniel thanked John and promised to check with him later for the end of the story. According to Daniel, "John's story shows an interesting case of robots, where standardization should definitely be implemented." Another trainee also mentioned that in their plant, operators usually reorganize their workstation before starting production.

Daniel carried on: "In those two examples, there are two kinds of losses. They both look like self-inflicted wounds. In effect, those changes at the beginning of each shift take time, just like a tool or die change we are imposing on ourselves. The second loss comes from the loss of opportunity to leverage other teams' or operators' good practices. The latter is our focal point." Daniel stopped for a deep breath, then asked if anyone had questions so far. Facing no response, he continued: "Now, let's go back to the simulation exercise we have been using from the beginning."

Over his years of giving trainings, if there was something Daniel had learned was powerful, it was the importance of preparing the trainees. He wanted to make sure up front that people were in the most open-minded disposition. As he used to say: "To ensure that people open their head's door," he frequently used quizzes. He clearly knew that there was no better way to capture people's interest and prepare them for learning than an anodyne question. Quite unsurprisingly, the one he chose looked trivial. "Before we start working on the simulation, could you guess the percentage of improvement we would be able to achieve by applying 'black book sharing' to our simulation? Are we talking about 5%, 10%, 15%, 20%, or more?" The room stayed silent for a few seconds. Then, Thomas reacted: "Well, I really have no idea; however, based on my experience running plants, I think getting a 10% improvement through any project is not so bad. Therefore, I would take 10% if you can give it to me, Daniel." Daniel nodded and continued: "Well, 10% for Thomas. Who else wants to play the guessing game?" Many attendees responded.

Daniel wrote down all numbers, including the boldest, 30%, given by Eric, who, as a smart good risk taker, sensed that the interest for Daniel asking such question would only be justified by a surprising outcome. Most of the numbers ranged from 5% to 20%. Daniel then concluded: "I have written down all your numbers, let's leave them here. We will see who got the best guess in a few moments." John, the engineering manager, jumped in and argued that the real improvement he saw was on the machine; he went on to say, "I have noticed that the operator is waiting while the machine is ironing. This is slowing the work; ergo, if we can speed up the machine a little bit, we will get some real improvement." This point of view from an engineering manager came as no surprise to Daniel. "They keep focusing on machine capacity. It is absolutely Pavlovian; when they hear improvement, they think machine," Daniel mumbled to himself. In effect, John was so focused on the machine that he did not even notice that his comment was out of the context of the question. Daniel responded: "Well, John, I'll take note of your comment, but I suggest we come back to that point later, when we will be talking about ways to improve the process, all right?" John acquiesced, visibly a little embarrassed.

Daniel then moved to the paperboard and started flipping the charts back. He stopped on one* (Figure 4.1). "You surely remember this table that we filled out yesterday: Sixteen numbers, each recording the times of the four steps of the four teams. You remember when we took the mode of each of the four steps of the packing operation: picking the T-shirt, ironing the T-shirt, folding the T-shirt, and storing the T-shirt, right?" The room acquiesced. "Well, along with the time, I made some notes assessing your work, quality-wise and safety-wise. I used those numbers to build another table," Daniel said, pointing to a chart he had displayed early this morning while preparing the training (Figure 4.2).

"We will use this benchmark table to share the best practices between teams. Let me underscore this again. As you may notice, it is not only about time. We are also interested in safety/ergonomics and quality. Well, let me state this very strongly: No time-saving good practice should be retained if it hampers safety or quality." This was certainly not the first time Daniel was insisting on the primacy of safety and quality over time

---

* Excerpt from the previous book: *Implementing Standardized Work: Writing Standardized Work Forms.*

	Team 1	Team 2	Team 3	Team 4
Picking the T-shirt	7	21	5	21
Ironing the T-Shirt	11	11	11	12
Folding the T-shirt	11	24	19	28
Storing the T-shirt	8	5	7	11
Cycle Time	37	61	42	67

FIGURE 4.1

Modes of the four measurement steps of the four teams carried over from yesterday's session; more details are available in the previous book of the series: *Implementing Standardized Work: Writing Standardized Work Forms.*

saving. It was therefore no surprise to the trainees, who murmured their approval. "Well, coming back to the time," Daniel carried on, "if you look at the 'Total' row and column 'Best,' you will see that there are two numbers. The first one, 37, is the best cycle time of all four teams and the second one, 32, is the total of minimum times for each step." Daniel paused a few seconds to make sure that everyone was following so far, then continued. "Now I have a question for you: Which conclusion do you draw from the gap between those two numbers?" Eric reacted instantly. "The fact that the

		Team 1	Team 2	Team 3	Team 4	Best	Reasons for being the best?
Time	Step 1	7	21	5	21	5	
	Step 2	11	11	11	12	11	
	Step 3	11	24	19	28	11	
	Step 4	8	5	7	11	5	
	Total	37	61	42	67	32 / 37	
Safety/ Ergonomics Issues			1	0	2	0	0	
Quality Issues			1	2	0	3	0	

FIGURE 4.2

Benchmark table for sharing the best practices between teams.

total of minimums is 5 seconds below the minimum of cycle times proves that the best operator is not always the best at all steps. Therefore, grasping the method of the best team to make it the new standard will not always be enough. We need to go into more detail to fetch the best for each step and spread it to everyone else." Daniel thanked Eric and added:

> What you just said is completely correct. I actually built the new table you just saw for this purpose [Figure 4.2]. Okay, as it is now clear to everyone, we will examine every step, identify the best team, and find out the reasons why the best team is actually the best. Standardized Work forms you all know support this quest for best practices: the Standardized Work combination table, Standardized Work chart, operator work instructions, and Process analysis chart. In practice, that means that you will have to compare those documents between the four teams. Also, when you go to the shop floor, shooting videos can be very useful. Actually, displaying different operators' work side by side in a room with the whole group around will make it easier to see different practices and thereafter identify the best ones.

Daniel moved to the paperboard and started a table while explaining (Figure 4.3):

> Now, before I let you do the black book sharing, let's make some quick computations to look at potential savings in two cases: when you only copy the best team and when you copy the best from within each team. First, here

	Team 1	Team 2	Team 3	Team 4	Total 4 teams
Current Cycle time (sec)	37	61	42	67	12,2
Current production rate (parts/min)	1,62	0,98	1,43	0,90	4,93
1-All teams at 37 sec (parts/min)	1,62	1,62	1,62	1,62	6,49
Percent improvement	0%	65%	14%	81%	**32%**
2-All teams at 32 sec (parts/min)	1,88	1,88	1,88	1,88	7,50
Percent improvement	16%	91%	31%	109%	**52%**

FIGURE 4.3
Potential saving from black book sharing.

are the cycle times of all four teams. Now, you can trust me, guys, the best way to navigate safely in this thing is to work with production rates, not cycle times. Therefore, I will write down in the second row the current production rate of each team. This is given in parts per minute. Hypothetically, let us suppose that we only copy the method of the best team: Team 1. That means all teams will be at Team 1's cycle time. Overall, this will result in a 32% increase in production. If we can copy the best of all steps of all four teams, then the potential for improvement increases drastically. It reaches a staggering 52%! Hey, folks, think about this—we are talking about gaining 52% just by looking around at what the other guys are doing! The beauty here is that this bounty comes at virtually no cost. This isn't a big CAPEX action. It's zero CAPEX!"

Thomas jumped in and asked: "Daniel, I am really surprised that the gain could be so significant. Is this the order of magnitude you really notice in the plants that have implemented Standardized Work, or are you just making it huge in this simulation to show the importance of black book sharing?" Daniel, who was about to address this point, was visibly pleased that somebody asked the question:

Well, Thomas, this is an excellent question. I can tell you that I have seen even more gains in real plants. The minimum I have seen so far is 5% in one of our German plants, actually one of our top plants, and the maximum was 75% in a plant in the United Kingdom. Most of the numbers have been between those two numbers. As you can see, it is a really big deal, yet people tend to overlook such opportunities. For instance, in the U.K. plant, nobody actually knew that one of the operators working on weekends was 75% more productive than the others. When this guy heard that the plant was launching a Standardized Work initiative that would include sharing best practices, he voluntarily showed up and explained that he had a method he figured out alone that, he believed, was more productive than others' methods. Believe me, this is a true story. Now, think about all those years the plant had troubles delivering its big-margin products to its customers while it could have hugely increased its capacity simply by copying the method of the weekend operator!

After this appalling story, Daniel continued: "OK, guys, I will be very gracious with you. I will not go back to check your initial improvement guess. I think Eric gave the best guest. He anticipated 30%, 22% shy of our final result, which is 52%."

Reacting to an attendee's remark, Daniel explained that they needed to get beyond the numbers and see concretely how the 52% improvement could be achieved. That would ultimately be done using Standardized Work forms and videos, which he lauded as an excellent tool to perform comparisons between workstations. Daniel proposed that Eric facilitate the benchmark between teams to reveal all the good practices and write them down in the last column of the table he taped on the wall (Figure 4.2). He insisted that the group must not go beyond his instructions to find more improvement. "This is the first phase of the improvement process, which should be limited only to black book sharing. Please do not try to find more improvements at this stage. As discussed previously, almost all actions you will find here will fall into the category of what I call 'just-do-it.' These are actions whose implementation requires relatively little effort and yields important benefits," he reminded.

After a few minutes of discussion, the group filled out the table (Figure 4.4) and listed the good practices collected on a flip chart (Figure 4.5). Almost all of them could be implemented quickly. Therefore, Daniel asked each team to apply them right away. A few minutes later, the implementation was completed. Daniel asked each team to use a stopwatch to evaluate the new cycle time. Team 1, still the best, reached the 32 seconds set as target.

		Team 1	Team 2	Team 3	Team 4	Best	Reasons for being the best?
Time	Step 1	7	21	(5)	21	5	Team 3's Raw material bin closer to station
	Step 2	11	11	11	12	11	No difference. Press time very stable
	Step 3	(11)	24	19	28	11	Team 1's operator folds faster, has better dexterity, lays down the T-shirt flatly before starting.
	Step 4	8	(5)	7	11	5	Team 2's finished goods bin is closer
	Total	(37)	61	42	67	32 / 37	With 37 sec, Team 1 is the best but could reach 32 sec if it was the best at each step.
Safety/ Ergonomics Issues		1	0	2	0	0	Team 2, 4's hands always kept out of the press. Operator's path cleared.
Quality Issues		1	2	0	3	0	All Team 3's folded T-shirts respected quality requirements

FIGURE 4.4

Cycle time, safety, and quality of four teams show that the overall best team was not the best in everything.

Good Practice-Sharing Actions
- Move raw material bin close to the workstation (learned from Team 3).
- Copy Team 1's folding method; make sure that it ensures a good-quality product.
- Move finished goods bin close to station (learned from Team 2).
- Inform/train operators to keep hands clearly out of the press (learned from Team 2).
- Clear operator pathway (learned from Team 2).

FIGURE 4.5
Main actions from black book sharing.

One point appeared not so obvious to implement: copying Team 1's folding method. Daniel explained that the performance of Team 1's operator, as stated in the column "Reasons for Being the Best?" of the chart in Figure 4.4, was a combination of the pace of work, dexterity, and a key point, which consisted of laying the T-shirt flat before starting. He then summarized his message in a table (Figure 4.6) in a few words while explaining:

As you all know by now, increasing the pace of the worker by making the worker run is not the objective of Standardized Work. Once again, we focus on the best way to do things, which will translate into more productivity and increased safety and quality. The goal is not to get people to work harder but smarter. Now, let me jump to the third item, which is a key point: laying the T-shirt flat before starting. This is exactly what we are looking for. This is something clearly identified that will be easy to spread, first through Standardized Work documents and later in training documents as well. Now, let's move on to the last and trickiest item, which is dexterity. This is a combination of a lot of things. The least I can say is that it takes time to acquire dexterity when it is not natural. Therefore, it should be addressed early when hiring operators. Then later, intensive training or

Reasons for being the best	Copying	How to address?
Pace of work	No	-Not the focus point of Standardized Work. -Standardized Work is about the best way, not the pace.
Dexterity	Possible but not easy	-Screening when hiring operators -Intensive initial training -Increases over time
Key point: "lay T-shirt flat before starting"	Yes	-Include as key point in Standardized Work documents -Include in Standardized Work training documents

FIGURE 4.6
Difficulty in copying Team 1's way of folding T-shirts.

experience accumulated over time can improve things. I do not want to spend too much time on this topic now. We will address it tomorrow when we talk about training.* The bottom line here is that it is normal that the operators of the three other teams do not get the 11 seconds of Teams 1's operator instantly.

One participant wanted to know if the next step would be to update the Standardized Work forms. Daniel, with a favorite line, "If the student has not learnt, then the teacher has not taught," was an extremely patient teacher. He took this opportunity to reexplain what he had already said: "As I have explained previously, the updating step will come after the next wave of improvement, called *Tachinbo*, prioritization and execution. And—your question comes at the right time: Tachinbo is exactly what we are going to do next."

Another participant had a question: "Daniel, you have underscored several times that Standardized Work is not about time, yet the first thing you do is to compare the modes of the four operators. I am a little bit confused. Could you explain?" Daniel thanked the participant for the question, which was an opportunity to clarify the methodology he just presented. "Well, look: As you may have noticed, we have ended up with proposed actions to improve each operator's method. I mean that we used the time to identify the possible best operators, then we asked ourselves reasons for these individuals being the best, which at the end of the process led to a list of suggestions for improvement actions." He talked while moving toward two charts (Figure 4.4, Figure 4.5). "As the saying goes: 'The time is the shadow of the method,' and we are engaged in a sort of inquiry in which time is used as means to evidence discrepancies between the four operators' production methods. It is actually difficult, and almost impossible, to find two operators using different methods that result in equal times." Daniel then drew a simple table to illustrate his words (Figure 4.7) and concluded: "The operator with the minimum time has the highest probability to possess the best method. We then need to make sure that this time is reached in the smartest—not the hardest—way that does not hamper safety and ensures a good-quality product."

Daniel congratulated the group and asked Thomas if he would like to summarize the learning before they moved to the next module. Thomas went to the paperboard and laid down the takeaways (Figure 4.8). Daniel,

---

* Training is the subject of the next book in the same series.

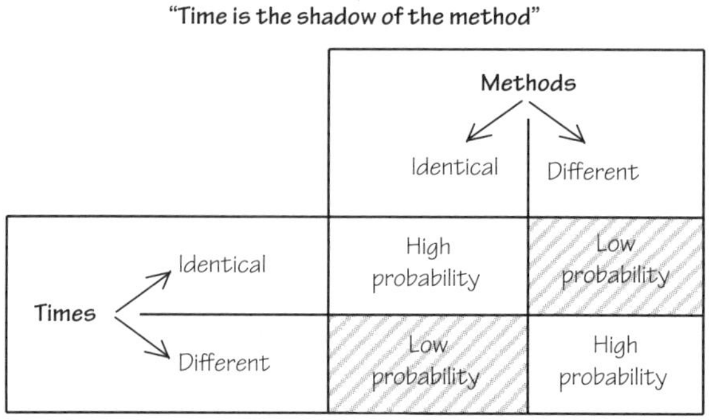

"Time is the shadow of the method"

FIGURE 4.7
Different times are strong evidence of different methods.

Black Book Sharing Takeaways
- It consists of sharing all operators' good practices.
- It should not be limited to the operator with the best cycle time.
- Rather, look at the best operator for each step.
- It should go beyond time to include safety, ergonomics, and quality.
- No time-saving "good practice" should be retained if it hampers safety or quality.
- Most black book sharing resulting actions are "just-do-it" actions—very low effort with substantial benefits.

FIGURE 4.8
Main black book sharing takeaways.

obviously satisfied with those bullet points, found nothing to add. He therefore suggested they move to the next part, dedicated to Tachinbo, after a 5-minute break.

> The best operator is not always the best at every step. Therefore, grasping the method of the best team to make it the new standard will not be enough. We need to go into more details to fetch the best at each step and spread it to everyone else.

> Almost all actions you will find here will fall into the category of what I call just-do-it. These are actions whose implementation requires relatively little effort and yields potentially important benefits.

5

Tachinbo

During the 5-minute break, Daniel had a quick discussion with Thomas. Daniel had, by chance, heard that there was a project under way to lay off several workers, and he wanted to verify the information. He commented: "You certainly remember the discussion we had before launching Standardized Work implementation. I cautioned that everyone must be on board, workers first. No operator should think that he or she might lose his or her job as a result of improvements. Look, fundamentally, I have nothing against cutting workforce when it is necessary, but if it needs to be done, it has to be done plainly and at the beginning. Afterward, a commitment should be made to the remaining workers to keep their jobs safe." Thomas denied the rumors: "I have given them my word that no one will be fired." He admitted that there was actually a plan in the works but not layoffs. He promised to give more information to Daniel later. He also added that he ought to communicate quickly on the point to avoid those rumors becoming a distraction at the very moment he needed his entire staff to be focused on Standardized Work implementation. The break was now over, and Daniel had to carry on with the training.

Daniel expected this fourth day of the training to be very tough, yet he saw no sign of weariness on the mostly happy faces when the group returned from the short break. Flattered by this encouraging sign, he went directly to the heart of the matter; directing his laser pointer to the word *Tachinbo*, he started. "In this phase, we will focus on waste elimination. Well, before going any further, I need to give you the meaning of this Japanese word. It means standing still, and in our business, it refers to the act of standing before a process and observing it for hours. It is recounted that this was Taiichi Ohno's[*] favorite training exercise. He would

---

[*] Ohno was a Toyota engineer who is considered to be the father of the Toyota Production System or Lean manufacturing.

Muda = Waste
Mura = Variability
Muri = Overburden

FIGURE 5.1
Muda, Mura, and Muri must be eliminated to improve the process.

reportedly draw a circle and let his student stand inside; Ohno would leave and come back roughly 8 hours later and ask: 'What did you see?'" A diffused rumble arose from the room, signaling the trainees' astonishment. "What could one be looking at for 8 hours? Isn't that a waste of time?" they asked. Daniel expected such reactions. Also, he knew that until people learned by doing themselves, they would not understand any explanation. At this specific moment, he thought of one of his favorite quotations from John Keats: "Nothing ever becomes real til it is experienced." He therefore responded: "Well, I suggest we come back to those questions at the end of this training. Trust me, before you experience it, it will be difficult to understand. However, be assured, I will not ask you to stay on the shop floor today for 8 hours." He concluded jokingly. "Now, if you do not mind, I suggest that we go straight to the heart of the subject. Shall we?"

Daniel moved close to a chart and started a drawing while explaining (Figure 5.1). "There are three kinds of wastes according to Toyota: Muda, which refers to pure waste; Mura, which refers to variability or unevenness; and Muri, which refers to overburden of machine or, more importantly, people. It is often referred to as the 3 *M*s or the 3 Mus. This can be represented as a three-legged stool. Okay, when we stand still in the plant, we will be trying to identify those kinds of wastes." At this point, Daniel wanted to reiterate a few obvious things, starting with Muda. "As you all probably know, there are seven types of Muda that you can easily remember with the following mnemonic method: T-I-M W-O-O-D." Daniel went to the flip chart and started explaining (Figure 5.2). "T for transportation, I for inventory, M for motion, W for wait and not waste as commonly mistaken, O for overproduction, O for overprocess, and D for defect." He then underscored: "Of the 3 *M*s or

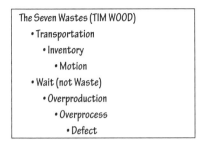

FIGURE 5.2
T-I-M W-O-O-D is the most common mnemonic to remember the seven wastes.

Mus, Muda is the most diffuse. However, it is also the easiest waste to see. To make sustainable change, it should be addressed after Muri and Mura, as you can see on the chart I drew this morning [Figure 5.3]." A trainee interrupted Daniel and asked him if he could elaborate a little bit more. Daniel replied, "I mean, it is very easy to see Muda. Even for a beginner, a few cycle times will suffice. When it comes to Mura, you need a much longer observation, as well as more education. Identifying Muri, I mean ergonomic strain, which needs both more knowledge and more sophisticated tools. Talking about implementation, I will simply repeat myself again: To reduce Muda in a sustainable way, you need to take care of Muri and Mura first.* We will discuss this point again shortly. This is basically the message of this chart [Figure 5.3]."

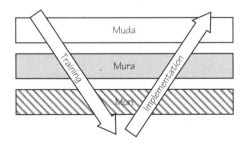

FIGURE 5.3
Muda is the easiest waste to see or train for, but it should be addressed after Muri and Mura.

---

* For more details, refer to the first and the second books of the series, respectively: *Implementing Standardized Work: Measuring Operators' Performance* and *Implementing Standardized Work: Writing Standardized Work Forms*.

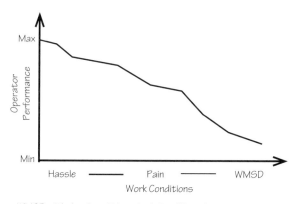

WMSD = Work-related Musculoskeletal Disorders

FIGURE 5.4
Operator performance is affected as the operator begins to feel the effects of an ill-designed workstation.

Daniel felt the time was right to underscore the key objectives once again:

Please remember what our key objectives are: reaching waste-free and stable operator work. Naturally, we will be focusing on the operator work method: his or her motion, following his or her legs, feet, and arms. Try to see if there is waste that can be eliminated by suppressing some movement or arranging the sequence differently. It is mostly about eliminating Muda here. Now, as we saw a few minutes ago, for the Muda to be removed in a sustainable way, we need to eliminate external causes of unevenness, or Mura. The main one comes from the part-feeding systems, equipment, or machine. Besides part-feeding bins or systems, we also ought to focus on machine problems. The main ones and the most visible ones are breakdowns, stoppages, and changeovers. I will come back to this specific point regarding machines later. Another thing we need to be focusing on to sustain waste reduction is Muri. The biggest impact of Muri is when the operator is overburdened. Therefore, we will be looking to improve work conditions, which are mostly ergonomics, postures, and unsafe behaviors. As you may see on the chart that I just drew, work conditions have a big impact on worker performance* [Figure 5.4].

---

* Although the curve is a rough representation or the evolution of worker performance as a function of the work conditions, it is largely supported by several research works, including the ones by H. Monod. Also, data collected by Alain Patchong, the author, and his colleagues have shown that there is a nearly linear correlation between an operator performance and his or her work conditions.

The chart in Figure 5.4 shows the performance of an operator decreasing as work conditions degrade. Daniel explained that the most important point was not the details of each bump of the curve, but its trend depicting operator performance versus work conditions. He then went on to add:

> I would like to take this opportunity to insist on an often-ignored fact: "Work doesn't need to be a pain!" Degraded work conditions hinder an operator's performance. It starts with hassle, then becomes pain, and ends up as an injury. As the process continues, the worker's body performs poorer and poorer until full stoppage, when injury arrives. As our regional HSE [health, safety, and environment] manager always says: "What beats your people beats your rate!" There is a clear link here. Therefore, remember that every time the company improves worker conditions, it is neither lost money nor a philanthropist deed. It's a win–win. It's a win for the worker who can avoid WMSDs [work-related musculoskeletal disorders] and enjoy a good life, and it's a win for the company, whose productivity increases.

Daniel paused for a few seconds, then moved to the flip chart. It was time to refocus on Tachinbo's key points.

> Let me summarize our mental process so far [Figure 5.5]. Remember our initial objective was improvement. To improve, we need to remove waste.

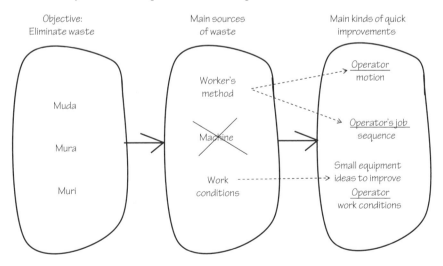

FIGURE 5.5
Standardized Work Tachinbo results must be related to worker motion, job sequence, and small equipment ideas. The focus is on the operator.

I told you that we had three kinds of wastes: Muda, Mura, and Muri. Remember the three-legged stool. We then saw that there were three main sources for those wastes: worker methods, work conditions, and machine stoppages. This is the spirit of Tachinbo in general. Now, remember that we are using Tachinbo here within the framework of Standardized Work implementation to capture quick, simple, zero-CAPEX improvements, and ideally, the ones directly related to the operator. Fixing machine problems will mostly need deeper analysis and more time to implement. While we are on the shop floor, we can take note of these points and communicate them to the plant's engineering department. However, these points are clearly not our target.

Now, because we are focusing on quick actions, the type of fixes we should be looking at are the ones related to operators' motions, operators' job sequences, and small equipment ideas to improve operators' work conditions. The question is how to tactically achieve those three kinds of improvement cheaply. Well, there are four cheap actions that can be implemented. As you can see on the chart [Figure 5.6], it's not like "one action, one improvement." For instance, one of these types of actions will have an impact on two kinds of improvements. In effect, a change in task sequence, in combination with Standardized Work in Process [SWIP], which will by

		Main kinds of quick improvements		
		Operator motion	Operator's job sequence	Small equipment ideas to improve Operator work conditions
4 Cheapest ways to improve	Change task sequence, combination, and Standardized Work in Process	×	×	
	Compact layout to reduce motion	×		
	Simple equipment improvements that enhance parts supply to improve operators' work conditions			×
	Simple equipment & ideas improvement to reduce time, improve safety/ergonomics and quality…	×	×	×

FIGURE 5.6

The four cheapest actions to improve operator motion, job sequence, and work conditions.

What beats your people beats your rate! There is a clear link here. Therefore, remember that every time the company improves worker conditions, it is neither lost money nor a philanthropist deed. It's a win–win. It's a win for the worker who can avoid WMSDs and enjoy a good life, and it's a win for the company, whose productivity increases.

definition improve the operator job sequence, will also help to reduce his or her motion. Conversely, to improve work conditions, one should chiefly focus on material supply at the workstation, which, in general, is the biggest source of work condition degradation. However, other small equipment ideas unrelated to material feeding could improve work conditions as well. Okay, those four cheap actions [Figure 5.6] are definitely the kinds of actions that can be implemented quickly with little effort—I will even say that they can be carried out within hours.

After the proposed action, estimating the benefit of each proposal is the next step, Daniel explained. A participant raised her hand and made the following remark: "Daniel, I do not see the point of wasting time on an estimate that is probably wrong." Daniel thanked the person for the opportunity to explain.

Well, the first reason for doing an estimate, whatever its accuracy, is that you need to base your selection or prioritization on something. The second, and most important, reason is that it will help you to learn a lot later. How? Well, we are clearly in the Plan-Do-Check-Act [PDCA] cycle. Coming up with your best estimate is the "Plan" phase of the PDCA; then will come the implementation phase (Do). You will be able to see the gap between your initial estimate and the result of the implementation; this is the "Check" phase. This gap is the best measure of your initial understanding of the physics of the process you are trying to improve. It shows how close you came to hitting your target. Ultimately, what comes next is the most important of all. Explaining this gap is a unique opportunity to learn and spread the knowledge, which leads to the "Act" phase of PDCA.

Now, let me tell you something. The best way to make an estimate is not to sit in your office and crunch questionable numbers. As always, and whenever it is possible, you will need to go to the shop floor and experiment or, more precisely, perform a mock-up. Use whatever you can find around you: cardboard, people—in a word, all assets available. Use them to emulate the real situation. Based on this mock-up, you will be able to achieve your best estimate. In general, there is rarely a better estimate. To support your communication during prioritization and later, take pictures and shoot videos during your mock-ups. Using video can be helpful as it allows you to clearly share your ideas with the rest of the group: Video of the current situation and video of the mock-up are used as the basis for the estimate. Videos are an excellent way to communicate your ideas.

Daniel had been talking for many minutes now. To relax the atmosphere a little bit, Daniel concluded with a German phrase well popularized by Thomas in the plant: "Alles klar?" The audience responded with laughter.

After those explanations, it was time to apply the learning to the simulation. Daniel asked each group to have their operator run a few cycles again while the other members observed. He handed out a few Post-its® and explained that each group would have 15 minutes to write down suggestions for improvement, agree on which ones they would like to display, and then stick them on the chart he unveiled (Figure 5.7). Daniel insisted that the group agreement at this level was merely a prescreening. The whole group would do final selection and prioritization later when all the teams came together. He asked the trainees to classify their suggestions in four columns corresponding to their impact: safety, quality, cost (mostly

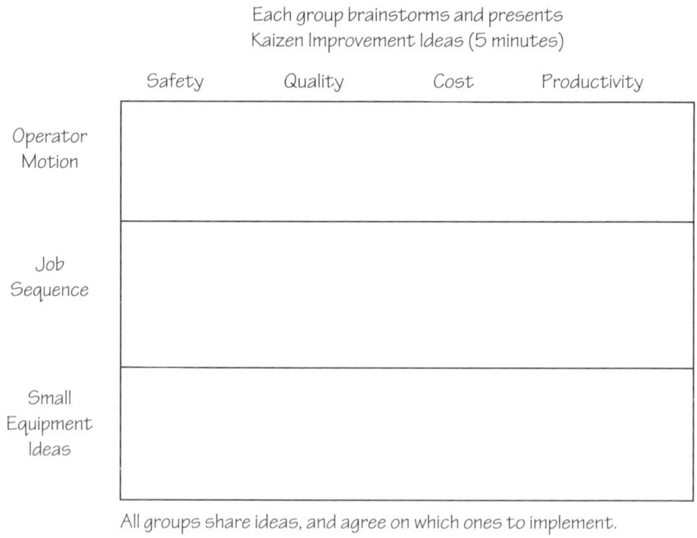

FIGURE 5.7

After agreement, each group's improvement ideas are written on Post-its and placed on the chart.

material waste reduction), and productivity. As someone remarked, a suggestion could have potential impacts in several areas; Daniel explained that they would have to place the suggestion in the area that fit the best, in other words, where it produced the biggest impact. He also reminded that proposals should be performed in two sequential phases: first, free suggestions, then selection of the best ones based on the group's estimate of benefit. "Please do not mix both; creativity first then selection next. This is the most efficient way I know," he stated. "If you do not follow this advice, you get yourself quickly stuck in arguments, and hours later you will still be fighting with some 'blood on the walls,'" Daniel concluded jokingly.

The task was now completed, and the dedicated chart was full of ideas from all teams. Daniel asked each team to come forward and share their proposals using videos. After all teams had shared their proposal, it was time for Daniel to conclude the Tachinbo session. He explained that this was what he had to say regarding Tachinbo in general. He then added, "I have developed some tools that can help implement Tachinbo in a very effective and focused way. I will share them with you later." He then suggested moving to the next module of the training. As the previous ritual goes, he asked Thomas if he would like to summarize. Thomas went to the paperboard and laid out a few bullet points about the main steps of Tachinbo (Figure 5.8). Daniel thanked Thomas before moving to the next part, dedicated to prioritization of listed actions.

Tachinbo = Intensive Observation

Goal: Find quick actions to reduce Muda, Mura, and Muri.

Steps:

Observation:

- Split in several groups.
- Each group makes observations on the shop floor in a specific area, identifying quick wins related to safety, quality, cost, and delivery.

    Cheapest actions: (1) compacting layout; (2) changing job combination sequence and SWIP; (3) improving material supply; (4) small equipment ideas to improve safety, ergonomics, quality, and productivity.

- One person per group videotapes/takes pictures.

Estimate benefit and effort:

- Each team estimates the benefit and effort of proposed actions.
- Mock up and take pictures and videos.

Sharing:

- Based on the estimate, each group selects which ideas should be shared with the others.
- Each group shares its proposals with the other groups using video to ease understanding.

Deliverable: List of actions.

FIGURE 5.8

Main steps of Tachinbo.

6

Prioritizing and Executing Actions

Thomas took advantage of the change of subject to have a quick discussion with Daniel. It was somewhat related to the discussion they had previously. He wanted to inform him that he would be leaving the training in an hour or two, with no certainty about the duration. He just had a phone call from his assistant, who informed him that the labor union had agreed to meet him this afternoon to discuss the future of the plant. He told Daniel that, for the time being, he could not give him more details but promised to share more with him later. Daniel thanked Thomas for letting him know, then refocused on the training.

Daniel started this module about prioritization with a quotation. "A famous author who knew Japan very well once said 'All good work is done the way ants do things, little by little.'* I think almost everyone in this room would agree with that. Now, the question is which 'little thing' do we start with? Well, this is where you need to be good at prioritizing." Over time, since his early days of giving trainings, Daniel had noticed that very often, after a workshop on the shop floor, people would end up with a list of actions and not know how to proceed, perhaps overwhelmed. Subsequent follow-ups would show little implementation of identified actions. Daniel therefore decided to include in the training a simple common tool for prioritization. It had received a positive reception over the years. It became very popular. Some plants even used it in their own workshops. As he always explained it: "Prioritization precedes action; it is the ground zero of any serious execution. Also, keep in mind that it must be a team effort, which thereby validates its highest commitment toward execution."

The prioritization tool Daniel talked about was a chart that had two axes (Figure 6.1). The horizontal axis represented the amount of effort to be

---

* Quotation from Lafcadio Hearn that can be read in *Simple Principles to Excel at Your Job* by Alex A. Lluch (WS Publishing Group, San Diego, CA, 2008).

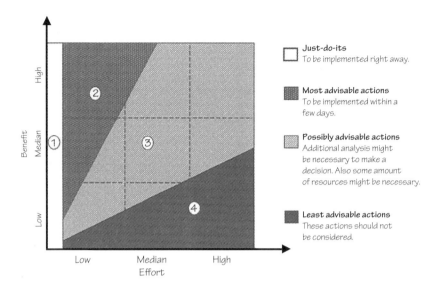

FIGURE 6.1
Charts to be used to set priorities.

invested in the implementation of the action. The vertical axis represented the benefit that could be derived from implementation of the same action. Both axes had three levels: low, medium, and high. On this chart, four areas were numbered from 1 to 4. Daniel explained:

> Area number one is dedicated to actions that request little effort. Benefit can be low to high. Because those actions need little effort, they must be implemented right away. Area number two is reserved for the most desirable actions. This refers to actions whose effort is low to mid-medium with a good level of benefit. Area number three includes possibly desirable actions. Effort and benefit are in the medium range. Most of those actions need some preparation. After the preparations, part of those actions will be implemented and the rest discarded due to their low benefit/effort ratio. The last portion of the chart, area number four, includes the least-advisable actions. Yielded benefits are low, while efforts are high. Those actions need, in general, some preparation or study activity to be validated and thereafter implemented. Unless the benefit is in the upper end, these actions are mostly scrapped.

After this explanation of the prioritization chart, Daniel suggested:

> I now propose we proceed to the simulation. It will basically consist of moving your Post-its from the previous chart [Figure 5.7] to the prioritization

Quick and cheap improvement	Typical priority	Implementation
Change task sequence, combination, and Standardized Work in Process to eliminate unnecessary motion and wait	1	Same day
Compact layout to reduce motion (walk and hands motion)	1 and 2	Same day to same week
Simple equipment improvements that enhance parts supply to improve operators' work conditions (variability, ergonomics)	2 and 3	Same week to same month
Simple equipment & improvement ideas to reduce time, improve safety/ergonomics and quality…	2 to 4	Same week to same quarter

FIGURE 6.2
Most common actions to improve quickly without huge investment.

chart [Figure 6.1] or the trash bin. Please place them based on benefits–effort according to the assessment made previously. Again, this is teamwork; therefore, we will be looking for consensus. Before I let you go ahead, I would like to quickly review the kinds of quick improvement actions that can be expected and their typical priority level.

Daniel displayed a chart he had prepared for the training (Figure 6.2). "Well, there are roughly four types of actions: the same ones we mentioned before [Figure 5.6]."

This time, Daniel wanted to be more specific than previously.

The first category is the kind of actions related to changes in task sequence, combination, and Standardized Work in Process to eliminate unnecessary motion and wait. These actions are the easiest ones. They can be implemented right away—I mean, at worst, the same day. These are just-do-its, which are ranked priority 1. The second type of actions consists of compacting layouts to reduce motion (walking and hand motion). These are typically priority 1 or 2. The third category includes simple equipment improvements that enhance parts supply to improve operators' work conditions (variability, ergonomics). Most of those actions can be implemented the same week. These are typically priority 2 and 3 actions. The fourth and last category is comprised of simple equipment improvements and ideas to reduce time or improve safety and quality, and everyone is free to propose these.

Daniel then concluded: "Now it is time to let you discuss the priorities of your actions." Before the prioritization started, Daniel had an important point to make, something the trainees had heard before, but was worth repeating once more: "No action should be considered if its implementation hampers safety or quality."

To nurture trainees' ownership, Daniel usually uses a local person to lead this type of exercise. "Eric, would you mind leading the group on this activity?" Eric happily accepted the role of facilitator.

The whole group agreed to a set of improvements and listed them in a table (Figure 6.3). The group selected seven ideas that they considered to be priority 1 and 2. Daniel was a little doubtful when it came to the discussion with the supplier to certify that it would be able to deliver 100% good-quality parts. Steve argued: "Normally, they should already be delivering 100% good parts. This is not like an additional service that we are asking from them. We will simply have to reiterate this requirement and enforce this request. At the beginning, we could set up a group to track and validate this point." Daniel, a strong believer in group commitment, caved in and let the group keep the action on its list. He underscored, to conclude the sequence about prioritization: "In the shop floor situation, the output of prioritization is an action plan, including among other information some key bits like the following: the person in charge, the committed start date, as well as the estimated date of completion and the status of the progress."

Daniel moved on to discuss implementation: "Implementation is really the first materialization of the reward for the group's work. Think about this: You are getting to the end of an intensive weeklong workshop, participants are tired, and some of them are starting to have doubts about what they are doing here—and voilà! Now they can see clear and concrete results of their work. Remember, these quick wins I mentioned earlier today are a big motivator." Daniel stopped for a few seconds to glance around, then carried on: "As I said before, be fast in implementation. No procrastination! Remember, as my father used to say, citing a famous person, 'The man who removes a mountain begins by carrying away small stones.'* Therefore, my advice is to go for imperfect execution with small gains rather than waiting for a perfect one with big gains that may never happen. If it is an improvement compared to the status quo and does not degrade safety and quality, then go for it." Daniel also explained that besides safety and quality checks, the gap between initial estimate and the actual result should be checked as well and eventually explained for learning purposes.

---

* William Faulkner's quotation that can be found in *SuperStar Selling: 12 Keys to Becoming a Sales Superstar* by Paul McCord (Morgan James Publishing, Garden City, NY 2008).

Quick and cheap improvement

Proposed actions

Job sequence, combination, & Standardized Work In Process (SWIP)

Eliminate wait: all wait = 0
= > inspect new t-shirt, Fold and inspect pressed t-shirt during auto cycle (SWIP = 2 in place of 1)

Motion/layout

Eliminate motion: all walk = 0
= > New very compact layout

Simple equipment (components supply)

Ergonomic material supply
= > Lift the table (work station) to achieve better ergonomic height (around 1.1 meters)

Simple equipment & improvement ideas to reduce time, improve safety/ergonomics, and quality....

Small equipment (+ programming)
= > to maintain the hot ironing press safely opened
= > Install 2 buttons on each side in place of one to close press safely.
= > Further improvement after closing press starts automatically
Simple ideas
No inspection of incoming T-shirt (customer quality certification)
= > detailed Step 2 eliminated*
no key points on inspection in major Step 1

FIGURE 6.3
Result of Tachinbo on the T-shirt-packing simulation.

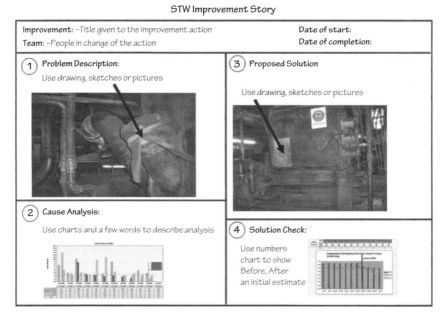

FIGURE 6.4

Standardized Work improvement story to be used for communication and progress reporting.

Daniel commented about the need to communicate about selected actions: "I am sure that the Excellence System guys have the right document that you guys can use. However, I would like to submit this one to you. It is a form that will help you tell the story of your improvement [Figure 6.4]." Daniel handed out to each group a few samples of a document, "Standardized Work Improvement Story." He explained that it had an oversimplified Plan-Do-Check-Act (PDCA) structure and included four parts: problem description, cause analysis, proposed solution, and solution check.

"The last point you do not want to forget," concluded Daniel, "is to celebrate. In the real situation, bring together the actors of the improvement and celebrate the success. The most important person around should chair the celebration—I mean the plant manager."

Before closing the session about prioritization and execution, Daniel proposed that Thomas make the summary. Thomas went to the paperboard and wrote down the main learning points on prioritization (Figure 6.5).

Key Points about Prioritization and Execution

- It is the ground zero of execution.
- It is teamwork, the first commitment of the team toward execution.
- Main steps:
  - *Classify* into four categories based on estimated benefit and effort: just-do-its, most advisable actions, possibly advisable actions, least-advisable actions. Use videos to share. Do mock-ups to estimate benefit.
  - *Plan:* Draw an action plan, including priority 1 to 3 actions, responsible person, and start/end dates. Scrap category 4 actions.
  - Implement the first category of actions the same day; category 2 should be taken care of within a week; category 3 must not exceed 3 weeks.
  - *Checks:*
    - Verify that implementation produces expected results (safety, quality, and cost).
    - For learning purposes, gaps between estimates (Tachinbo phase) and actual results should be explained.
  - *Communicate:* Use Standardized Work improvement story or any local document to communicate about action.
- Do something imperfect now; improve later.
- No action allowed if safety and quality are degraded.

FIGURE 6.5

Key learning about prioritization and execution: classify, plan, implement, check, and communicate.

After Thomas's summary, Daniel asked each group to get ready to run the T-shirt-packing simulation with an improved process and to take the time necessary to update the Standardized Work forms. This was the main subject of the next training module.

> Prioritization precedes action; it is the ground zero of any serious execution. Also, keep in mind that it must be teamwork. Most importantly, it is the first commitment of the team toward execution.

7

Updating Standardized Work Forms and Expanding

Daniel conceded:

> As you have probably noticed, I am a big fan of PDCA [Plan-Do-Check-Act]. Look, it is a very powerful method that increases the chances to achieve the best results wherever it is applied. Well, some of you have noticed that, but I want to acknowledge this to everyone. The training session we are now starting, "Updating Standardized Work Forms and Expanding," is, in effect, nothing else than the "Act" part of the PDCA approach applied to process improvement. I could have used the title "Acting" as well, but I found it less explicit.

After this short clarification, Daniel had an important message to pass on. "It is better to have a quick-and-dirty version of the document rather than none. So, do not leave this for tomorrow in order to make it perfect, just as mentioned previously for execution. Do it right away! Do it quick and dirty!" Daniel then moved to the middle of the room and asked the group to run the T-shirt-packing simulation again, take the needed time, and update their Standardized Work forms. Daniel had saved their previous forms. After the group completed the task, he asked the group to tape the "before" and "after" forms side by side so that the comparison could be obvious. As usual, Daniel interacted with the group to observe and answer questions. The work was completed, and he went back to the middle of the room again.

"It's now time to review your results," Daniel started. "Before we begin commenting on the results, let me settle a point John made earlier today about the speed of the machine being the biggest problem. Now, John, based on the result of your work, do you still think so?" John had no choice but to bow to the plain truth of the facts depicted on the Standardized Work combination chart (Figure 7.1): "It is clear that the machine time is not on the critical path of the cycle time; there is no doubt that it has no impact on the capacity of the overall process." Thomas jumped in and insisted on the need for everyone to base his or her judgment on real data coming from the *Gemba.** He also emphasized that his people should be focusing more on so-called zero-CAPEX improvement, which, he believed, could be found everywhere in the plant. He went on to conclude that this was exactly why he had decided to implement Standardized Work, which encompassed, among others, these two principles: the use of real data from the shop floor and a focus on zero-CAPEX improvement. Daniel thought there was no need to comment more on the point. He thanked John for providing this teachable moment to the group.

Daniel asked the group to comment on the differences between the before and after Standardized Work documents. Sarah, the human resources (HR) manager, who had been mostly silent, raised her hand and commented: "First of all, I am amazed by how much such a process can be improved. On Team 1, it looks big [Figure 7.1], never mind Team 4, that started with a cycle time of 67 seconds and is now running close to 17 seconds. That is really a huge improvement." Steve, the guy who knows the numbers, confirmed: "It should be around 75% improvement." Sarah then proceeded: "Thanks, Steve. I trust your numbers as always." Then, she refocused on her earlier point. "By working on two T-shirts during the same cycle time, we were able to eliminate the worker wait time. Also, we have compacted the workstation, as you may see on the Standardized Work chart [Figure 7.2]. A highly compact area put the worker at risk of body twist. To prevent operators from hurting themselves, we added a safety point to T-shirt picking (Figure 7.3)." Sarah breathed a few seconds and continued: "Yes, there is something else I wanted to mention. As we have discussed previously, we all agree 100% that quality checks before T-shirt folding are simply a waste if, by contract, our supplier is expected to provide us with good-quality products. This is something we should enforce even more in our plant." The plant quality manager acquiesced, which gave some encouragement to Sarah. "If they cannot deliver a good-quality product, they should pay a fine as agreed in the contract." Sarah concluded her intervention with comments on the Yamazumihyo charts (Figure 7.4). "The before and after Yamazumihyo

---

* *Gemba* is a Japanese word that refers to the shop floor, where production is done.

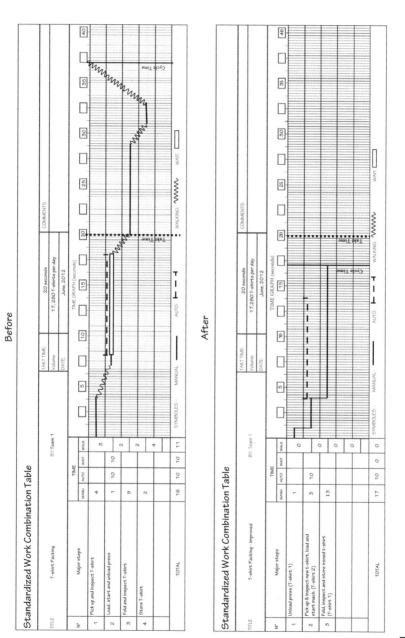

FIGURE 7.1
Before and after Standardized Work combination table: Processing two T-shirts (1 and 2) at the same time helps eliminate operator's wait time.

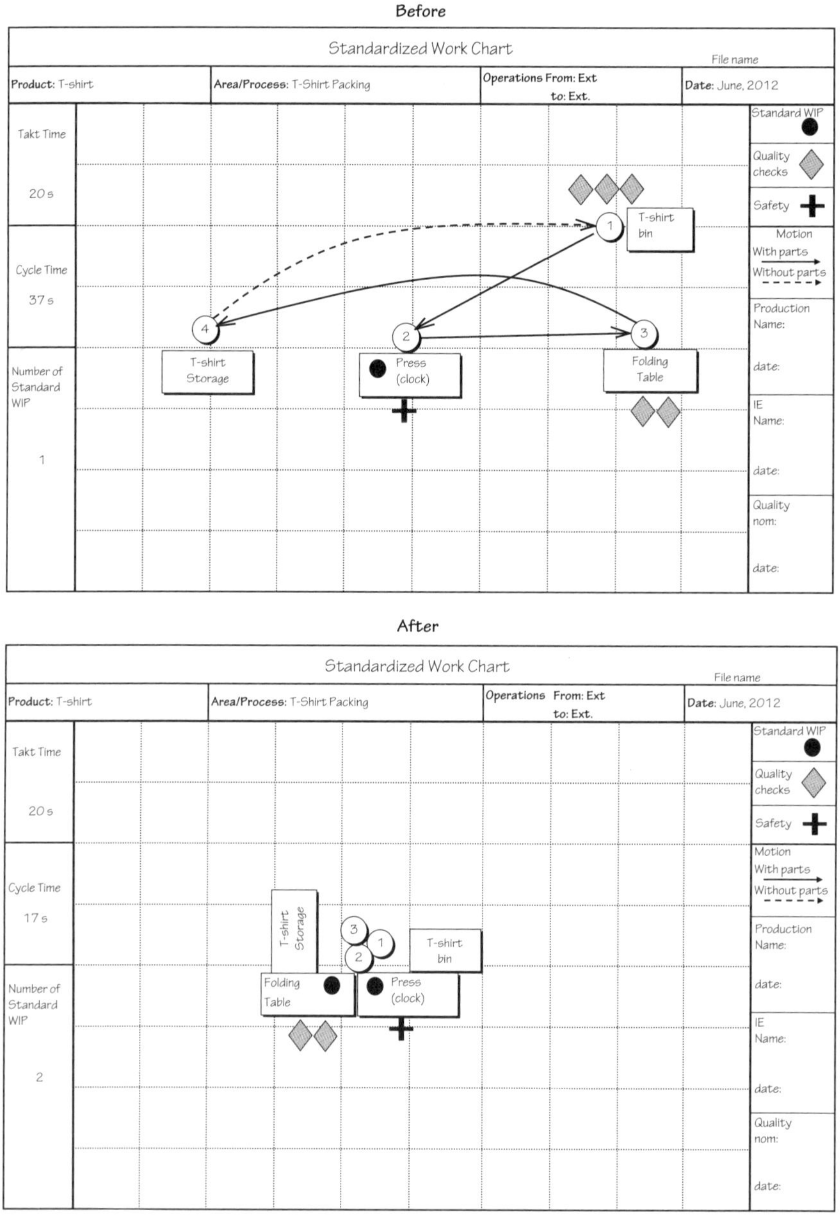

Before

Standardized Work Chart			File name

Product: T-shirt　　Area/Process: T-Shirt Packing　　Operations From: Ext to: Ext.　　Date: June, 2012

After

FIGURE 7.2
Before and after Standardized Work charts.

Before

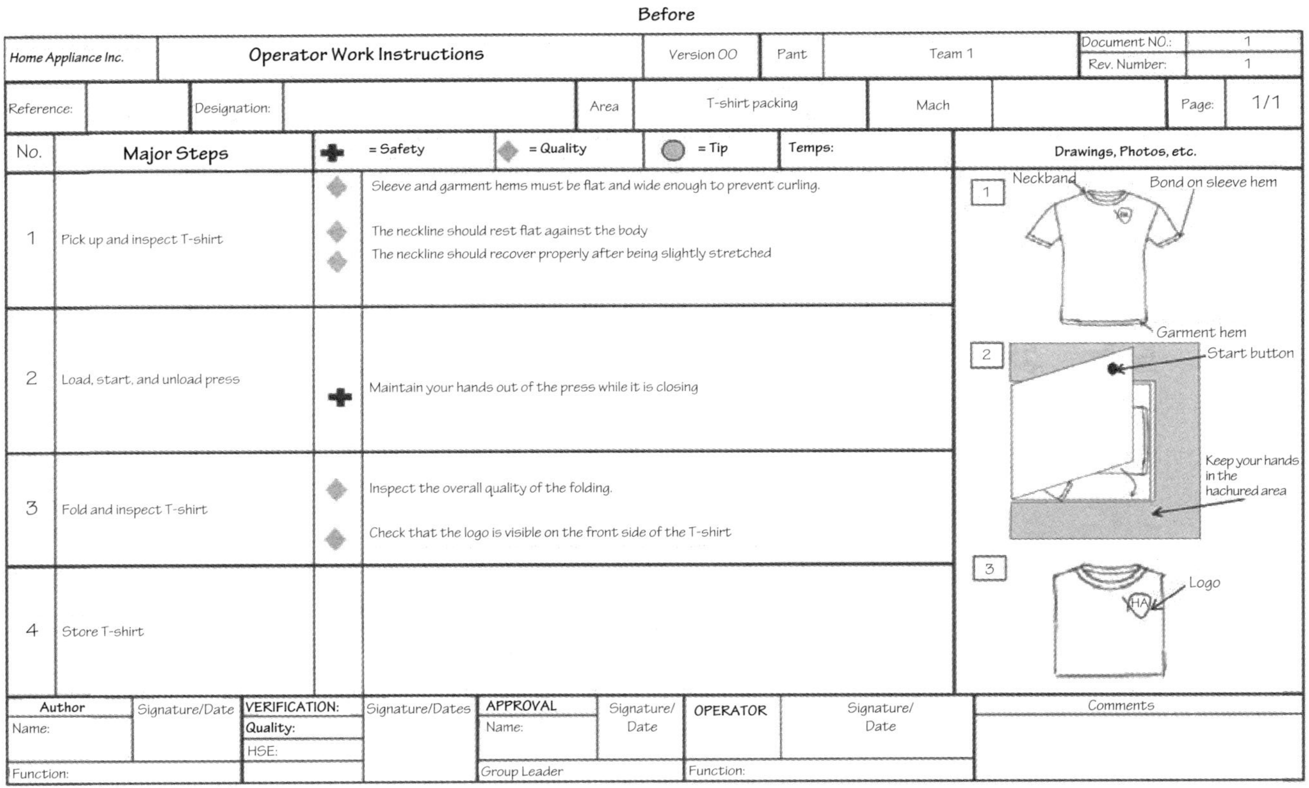

Home Appliance Inc.		Operator Work Instructions			Version 00	Pant		Team 1		Document NO:	1
										Rev. Number:	1
Reference:		Designation:		Area		T-shirt packing		Mach		Page:	1/1

| ✚ = Safety | ◆ = Quality | ⬤ = Tip | Temps: | | Drawings, Photos, etc. |

No.	Major Steps			
1	Pick up and inspect T-shirt	✚	*Sleeve and garment hems must be flat and wide enough to prevent curling.*	
		◆	The neckline should rest flat against the body	
		◆	The neckline should recover properly after being slightly stretched	
2	Load, start, and unload press	✚	Maintain your hands out of the press while it is closing	
3	Fold and inspect T-shirt	◆	Inspect the overall quality of the folding.	
		◆	Check that the logo is visible on the front side of the T-shirt	
4	Store T-shirt			

Drawings:
1 — Neckband / Bond on sleeve hem
2 — Garment hem / Start button
3 — Keep your hands in the hachured area / Logo

	Signature/Date	Signature/Dates		APPROVAL	Signature/ Date	OPERATOR	Signature/ Date	Comments
Author		VERIFICATION:		Name:		Function:		
Name:		Quality:		Group Leader				
Function:		HSE:						

FIGURE 7.3

Before and after Operator Work instructions. (Continued)

After

Home Appliance Inc.	Operator Work Instructions			Version OO	Area		Pant	Team 1		Document No:	1
										Rev. Number	1
Reference:	Designation:				T-shirt packing		Mach			Page:	1/1

No.	Major Steps	✚ = Safety	◆ = Quality	● = Tip	Temps:	Schema, Photos, etc.
1	Unload press T-shirts 1	✚				Start buttons — Keep your hand in the hachured area ☐ 2
2	Pick up & inspect new T-shirts, load and start Mach. T-shirts 2	✚	Rotate around right foot while keeping your body rigid			
3	Fold, inspect and store ironed T-shirts 1	◆ ◆	Inspect the overall quality of the folding. Check that the logo is visible on the front side of the T-shirt			Logo HA ☐ 3

VERIFICATION	Signature/ Date		APPROVAL	Signature/ Date	OPERATOR	Signature/Date	Comments
Quality:			Name:				
HSE:							

Author	Signature/ Date
Name:	
Function:	Group Leader: Function:

FIGURE 7.3 (CONTINUED)
Before and after Operator Work instructions.

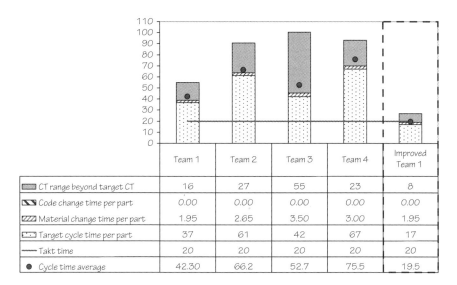

	Team 1	Team 2	Team 3	Team 4	Improved Team 1
CT range beyond target CT	16	27	55	23	8
Code change time per part	0.00	0.00	0.00	0.00	0.00
Material change time per part	1.95	2.65	3.50	3.00	1.95
Target cycle time per part	37	61	42	67	17
Takt time	20	20	20	20	20
Cycle time average	42.30	66.2	52.7	75.5	19.5

FIGURE 7.4
Before (all four teams) and after (Team 1) Yamazumihyo chart (time in seconds).

charts show a less-impressive reduction on the variability. However, the 3 seconds left between the target cycle time (17 seconds) and the Takt time (20 seconds) will be enough to absorb remaining variability. Well, that's all I wanted to say."

Daniel thanked Sarah and congratulated her for her substantial comments. John, still amazed by how the increase of SWIP[*] from one to two helped eliminate wait time, commented on this point. "As the 'After' Standardized Work combination chart depicts, the operator does not have to wait for the machine anymore; the operator can proceed with a value-adding task while the machine is running [Figure 7.1]. I clearly understand now why SWIP is said to be one of the four elements of Standardized Work. It is very defining—when it changes, everything else changes completely." Daniel added that increasing the SWIP was a common way to increase worker efficiency and sometimes machine utilization. He then asked if someone else wanted to comment further.

After a few seconds of silence, Daniel refocused his question. "All right, I remember that after we finished writing the Standardized Work documents, some of you asked me how they could be used to conduct

---

[*] SWIP stands for Standardized Work in Process and refers to the number of parts being processed. It is one of the four elements of Standardized Work. For more details, refer to the previous book of the series: *Implementing Standardized Work: Writing Standardized Work Forms.*

improvement.* Well, now that you have experienced their usage in process improvement, I want to return the question to you. How did you use those documents to foster your improvement ideas?" The first reaction came from Steve, the industrial engineering manager:

> The Standardized Work combination table is very interesting. At first glance, it is quite easy to identify what I would call the critical path of the cycle time. It is very clear whether a certain improvement will have an impact on the cycle time. In simple terms, if an activity is on the critical path, then its improvement will have an impact on the cycle time; otherwise, there will be no impact. It is also very easy to see the walk time, the wait time, as well as the combination of tasks between the operator and the machine. Wait and walk are obvious wastes that make sense to eliminate. Reduced wait means better manpower utilization. Less walk time may sometimes translate into better manpower utilization or additional wait time. In any case, it will always lead to improved ergonomics! The beauty here is that, as we saw earlier today, better ergonomics means better operator performance. So, at the end of the day, even when walk reduction is not on the so-called critical path of the cycle time, it still makes sense to reduce it as it will reduce operator fatigue and improve operator performance.

The next person to intervene was Eric. He chose to talk about the Standardized Work chart (Figure 7.2):

> It gives, at a glance, a clear idea about walk time and the layout in general. The number of crossings between arrows shows the amount of motion. As you told us earlier, the goal is to have the fewest crossing arrows as possible, something that is close to a C-shape, in an as-compact-as-possible layout. Here, for instance, we went from one crossing point in the before chart, to zero in the after chart.

Thomas then discussed how the Yamazumihyo was the perfect tool to see all variability, special cause, and common cause. "It shows where to focus first." He commented:

> The most important point about this chart is that you can see how your process is doing versus the Takt time. When the process cycle time is below

---

* This passage refers to the previous book of the series: *Implementing Standardized Work: Writing Standardized Work Forms.*

the Takt time, the chart will tell you if it is enough below to absorb current variability. It really helps answer the first question, the level of priority of the improvement, before launching any improvement. For sure, all improvements are valuable. But, since we do not have unlimited resources, we have to make choices from time to time. As a plant manager, this tool is just perfect for prioritization. Yes, it clearly helps identify which process to focus on first!

Daniel could not agree more.

Daniel concluded by mentioning the operator work instructions. "As you all know, key points are where the method is conserved. Remember, these are points that are critical to successful execution of the method by the worker! They are therefore highly useful, especially when you need to compare different operators or record best practices to be disseminated." Daniel summarized the points illustrated on a chart (Figure 7.5). This concluded the module on Standardized Work forms updating.

Daniel then moved to the other part of the session regarding expansion of improvement to other processes in the plant or elsewhere:

Use of Standardized Work forms in process improvement

Standardized Work Combination Table	Shows the critical path*
	Shows wait time**
	Shows walk time***
	Shows task combination****
Standardized Work Chart	Shows amount of walk***
	Shows wasteful job sequences or motions*****
	Shows spread of layout
Operator Work Instructions	Helps identify or describe key points
Yamazumihyo	Shows amplitude of special cause and common cause variation
	Shows the extra-capacity needed to absorb variability

*It helps identify actions with impact on the overall cycle time.

** Wait time reduction may not reduce overall cycle time but will improve manpower utilization.

*** In general, reduction will at best reduce the overall cycle time, and at least improve ergonomics.

****A better Combination Table and additional Standardized Work In Process can help reduce cycle time.

*****The best motion should be C-shaped, which has the minimum back and forth.

FIGURE 7.5
Standardized Work forms are critical tools of process improvement.

In a real-case situation, the focus up to this stage is only one, or a few, machines or workstations to avoid dispersing the team's energy. After updating Standardized Work forms, just like in any PDCA-like approach, you need to think about ways to extend validated improvements to other similar processes. Obviously, you should first start in your plant, then make sure that those improvements are also expanded to other plants. This generalization step belongs to zero-CAPEX actions. Look, this is clearly the kind of thing we are looking for. Think about it: The solution is implemented and validated; there is no risk! The only thing you need to do now is to simply "copy and paste" it somewhere else and reap the benefits. By the way, note that skipping this step amounts to creating a black book of your own, which is exactly what Standardized Work endeavors to eliminate in the first place. Remember what we said earlier today. Now, in practice, you should do it right away. If this is not possible the same day, then include this in the action plan we mentioned before—indeed, make sure that there is someone held responsible, with start and end dates.

It was now time to move to the shop floor. Before that, Daniel first needed to have Thomas sum up a few key points of the learning as per their ritual (Figure 7.6). Then, he explained that, before hitting the shop floor, he had something to share. This was a set of tools he had designed to support Tachinbo and make it easier for beginners.

Key Messages about Acting
- Better to have quick-and-dirty forms than none.
- Updated forms are the first materialization/reward of the group work—use them to celebrate.
- Everything else known about Standardized Work forms stands:
  - Standardized Work combination table shows cycle time-critical path (help define priorities), wait time, walk time.
  - Standardized Work chart shows motion through the number of crossing points and shows layout compactness, walk.
  - Operator work instructions document helps identify best practices to be disseminated.
  - Yamazumihyo chart shows variability to help define priorities.
  - Increasing Standardized Work in Process can help reduce manpower or machine utilization.
  - Reduced walk (even when it has no impact on the cycle time) improves worker peformance.
- Think about generalization:
  - Make sure that improvements are extended to similar processes (in the same plant first, then elsewhere).

FIGURE 7.6
Key learning about updating Standardized Work forms and generalization.

It is better to have a quick-and-dirty update of a document rather than none. So, do not leave this for tomorrow in order to make it perfect. Do it right away! Do it quick and dirty!

Skipping the generalization step amounts to creating a black book of your own, which is exactly what Standardized Work endeavors to eliminate in the first place.

8

Focused Tachinbo: Workstation Assessment

Thomas had left the room when Daniel started the module on workstation assessment. He had to attend the meeting with the labor union. Here, as in many European countries, he knew that if he were to succeed, he would need to make sure that, ideally, the union was on board, or at least not opposed to his actions. He had decided to go for a transparent and plain talk with the labor union. This was an unusual approach for a plant manager. His predecessors had mostly been all but candid. By any account, previous plant managers and the labor union had been playing a zero-sum game: If you win, I lose; if you lose, I win. Promises had been made but not kept on both sides. It was simply part of the business. Unfortunately, this led to a long history of mistrust and lost opportunities. Those self-inflicted injuries had long eroded the plant's integrity and contributed to the plant's current dire conditions. The EMEA (Europe, Middle East, and Africa) Manufacturing vice president would concede spontaneously: "We have made big mistakes of our own; we need to have a radically different approach toward labor unions."

Thomas was there to implement, among other things, this new approach, to build with the labor union a sense of "we are all in this together" and to promote a new motto: "We win together, or we lose together"—no more business as usual. For this fresh start, the regional senior management believed that Thomas was the right person. As a non-French individual with an international background, this would keep him away from a *priori* judgment. Ironically, some considered his background to be his biggest weakness. They thought that he would be "played like a puppet," adding: "He is not up to the labor union game." Regional senior management would equally admit the risk, and the biggest question in their minds was: "Could we play a different game with the same people who have been going the opposite way?" In the end, they believed that the urgency of the situation was leaving them with fewer and fewer options, which inclined them to take a chance and try something different.

As Thomas explained very candidly to Daniel on Wednesday evening during their long discussion: "You can see this change positively or negatively. The positive reading would be to say, 'Wow, senior management is thinking out of the box!' The less-encouraging way would be to refer to Anton Chekhov, who famously said: 'When a lot of remedies are suggested for a disease, that means it cannot be cured.' Is this encouraging, creative, out-of-the-box thinking or another of several remedies being tried on a desperate situation? Only the future will tell!" Whatever the rationale behind what Thomas called the "new doctrine," the good point was that it was completely in line with Thomas's character. Therefore, he would not have to force himself to follow through. More than anyone around, he was conscious of being on uncharted terrain and therefore needed to exercise some caution.

Before moving to the module on the shop floor, Daniel wanted to give participants a form that they could use to support Tachinbo on their workstations. He had noticed over the years that telling people to stand up in front of a workstation to observe did not work very well the first time. "French are not Japanese," he would say. "I cannot just do it Taiichi Ohno's way; people become bored quickly and lose focus." To avoid the failures he experienced at the beginning of the deployment of Standardized Work training in the company, Daniel had developed two forms to help Tachinbo practitioners keep focus. Those forms, which addressed the most common points of weakness of workstations, were based on factual, understandable, and measurable elements. As he explained: "Two people using those forms for an assessment should reach similar results." Because the forms were based on physical elements and served as guides to the user, they were very successful. Most of the people who used them "like the way it gives structure to guide the user to the right results step by step; you just have to follow the forms and the process." The other point he would make is that, "These tools assess material supply. The main reason for that, as I stated previously, is that in most industries, material supply is the biggest source of waste. I mean the source of creation of Muda, Mura, and Muri."

At this point, a participant interrupted Daniel with a question: "Daniel, please could you tell us what is different or new versus the quick improvement actions you mentioned earlier. I am talking about that chart over there [Figure 6.2]." Daniel went to the chart pointed out by the participant

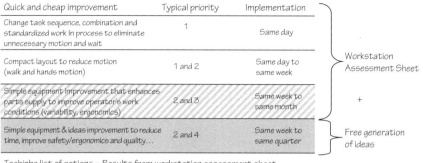

Quick and cheap improvement	Typical priority	Implementation	
Change task sequence, combination and standardized work in process to eliminate unnecessary motion and wait	1	Same day	Workstation Assessment Sheet
Compact layout to reduce motion (walk and hands motion)	1 and 2	Same day to same week	
Simple equipment improvement that enhances parts supply to improve operator's work conditions (variability, ergonomics)	2 and 3	Same week to same month	+
Simple equipment & ideas improvement to reduce time, improve safety/ergonomics and quality…	2 and 4	Same week to same quarter	Free generation of ideas

Tachinbo list of actions = Results from workstation assessment sheet
+
Free generation of ideas (mainly quality and safety)

FIGURE 8.1
The Tachinbo list of actions will come from a structured process based on a workstation assessment sheet and free generation of simple quick actions that improve safety, quality, and cost and delivery.

and explained: "Actually, the first three rows are addressed by the workstation assessment sheet. The last one, which I call free generation of ideas, is not included in the assessment sheet [Figure 8.1]. Since by definition free generation of ideas cannot be structured, I would say that the tools I will share with you take care of whatever can be structured." Daniel moved to a chart displayed on the wall and carried on: "This is the first form [Figure 8.2]. I have printed them out for you." Daniel handed out two sheets of paper to each participant.

Now, you all have two sheets of paper in your hands: The first one is an exact copy of the form depicted on this wall [Figure 8.2]. The second sheet [Figure 8.3] explains the main elements of the form. It is filled out to illustrate how it should be used: one per workstation. Now, let's go step by step. First, at each station you must list all components and tools requested in the operation of the station. Then, you will have to assess it for the three kinds of waste. I have selected the most frequent type of Mura, Muri, and Muda, which I call "hit criteria." As you can see on the chart [Figure 8.4], the posture, number, position, and changes are the hit criteria related to Mura. They define the four main invariants that must be achieved to reduce part-picking-related variability: invariant posture, invariant number, invariant position, and invariant time. This is what I call the "4Is." The acid test is that you need to be able to close your eyes and pick up exactly the part you need, when you need it. In other words, parts supply is deemed optimized when a blind person can perform the picking. 4Is are illustrated on this paper I am handing out [Figure 8.5].

Workstation's components or tool	Posture a	Number b	Position c	Changes d	Distance e	Height f	Body contact g	Twist h	Two-hand work	Walk	Wait	Main issue i	Current Measurement j
1 Plastic component	X	O	X	X	X	X	X	O				Distance/back feeding	100 cm
2 Metal structure	X	O	X	X	X	O	O	O				Distance/back feeding	110 cm
3 Safety device	O	O	O	O	X	O	O	O				Distance/back feeding	105 cm
4 Reinforcement sheet	X	O	X	O	X	O	X	O				Ergonomics	red
5 Rivets	O	X	X	O	O	X	O	O				Height	40 cm
6 Beads	O	O	X	O	X	O	X	O				Ergonomics	red
7 Hammer	X	O	O	O	O	X	O	O				Height	160 cm
8 Reveting tool	X	O	O	O	X	X	O	O				Distance	100 cm
9													
10													
11													
12													
13													
14													
15													
Final score	38%	88%	38%	67%	25%	50%	63%	100%	N	10	15		

Workstation Assessment Sheet

For each component or tool listed in the first column, fill in:
a "O" if operator's posture is always constant from cycle to cycle, "X" otherwise
b "O" if it easy to pickup the same number of components from cycle to cycle, "X" otherwise
c "O" if its position of the part is always constant from cycle to cycle, "X" otherwise
d "O" if all related periodical tasks (e.g., container change) are below 5% of the takt time, "X" otherwise
e "O" if distance to operator is below 30 cm, "X" otherwise
f "O" if the height of the location of component or tool is between 750 mm and 1200 mm, "X" otherwise
g "O" if operators has no body twist when grasping component or tool, "X" otherwise
h "O" if operators has no body contact when grasping component or tool "X" otherwise

FIGURE 8.2
Workstation assessment sheet based on eleven key criteria. (Continued)

Daniel handed out a sheet with four pictures to each trainee. "As always, a picture speaks more than a thousand words," he paraphrased. Before moving to the next part, Daniel felt the need to underscore, once more, the need to eliminate Mura. "Remember, I told you earlier when we discussed work conditions that 'Work doesn't need to be a pain!' Guess what, work doesn't need to be stressful either. I mean the operator should not be forced to think to avoid making errors every time he or she is performing a task. Everything should be natural…humm…automatic. Yes 'automatic' is the right word. Think about when you drive your car from home to your workplace every morning or the other way when you return in the afternoon. Most of the time you do not have to think… it requests minimum effort, not mental load. It is almost like a plane on autopilot. Well this

Workstation Assessment Sheet				
Improvement proposal	Measurement with proposed solution[k]	Gain (sec)[l]	Opportunities/gains[m]	Priority[n]
Kitting	30 cm	2	Ergonomics/Safety	
Kitting	30 cm	3	Ergonomics, variability	
Kitting	30 cm	2	Ergonomics/Safety	
Sequencing	green	2	Ergonomics, variability	
Move rivet bin up	80 cm	1	Ergonomics, variability	
New bead feeder	green	1	Variability	
Bring tool down	< 110 cm	1	Ergonomics, variability	
Movetool closer	30 cm	2	Ergonomics, variability	
Total gains (sec)[p]	14			

[i] Specify the main issue related to this component feeding
[j] Charaterize this main issue by a measurement (distance, time,...)
[k] Give your best estimate of the new measurement when improvement applied
[l] Give your best estimate gain is seconds when improvement applied
[m] List opportunities or non measurable gains that goes with improvement
[n] Give a priority to proposed action; from 1 to 4 based on the effort - benefit chart
[o] For each criterion in MURA and MURI column, fill in precent of "0" in the gray cell of "two hand" Coloum fill
-in "Y" if two-hand work is defined in forms and taught to workers
In the grey cell of "Wait" column fill-in the walk time per cycle
In the grey cell of "Walk" column fill-in the wait time per cycle
[p] Fill in the sum of time saved with proposed actions

FIGURE 8.2 (CONTINUED)
Workstation assessment sheet based on eleven key criteria.

should be the goal to achieve when we design an operator workstation. This reduces the disruption of the operator on the material flow and improves his or her work conditions as well." Daniel looked around as some attendees were nodding. "When it comes to workstation, less Mura also means less Muri," he concluded with a little smile. "Now, the next waste is Muri. There are four main work condition strains facing a worker when picking parts, which are my hit criteria distance to parts, height of part position, body contact, and body twist. Again, there are four drawings that illustrate each situation [Figure 8.6]." Responding to a participant's question, Daniel specified that the "distance to part" applied to whatever direction of the worker's movement is needed to reach the part: horizontal, vertical, or other. He also advised that the best way to avoid body twisting was to

MURA	Posture	Operator's posture must be constant from cycle to cycle. This point is about the change of posture, not a ergonomic assessment of the posture itself.
	Number	Operator should be able to easily pick up the same number of parts from cycle to cycle. This concerns mostly small parts (e.g., screws, bolts and nuts) or flimsy parts (e.g., harness)
	Position	The position of a part must be constant from cycle to cycle. This is not only the location of the part but also its orientation. In general, the Position varies if the Posture varies.
	Change	The change time for containers of parts feeding a device should below 5% of the Takt Time. This is not an average time per part but the whole time needed to setup a full container or feeding device when the previous one empties
MURI	Distance	Distance from operator to the part should be less than 30 cm This is a very challenging objective, however it drastically helps to improve the work conditions of the worker, thereby his or her efficiency.
	Height	Height of the position of a part should be between 750 mm and 1200 mm This ergonomic window normally depends on the gender of the operator and the weight of the part. The numbers given here correspond to a male worker and are related to part between 4 and 9 kg. If necessary, adapt the conditions accordingly.
	Body contact	No body contact must occur when picking up a part
	Body twist	No body twist must occur when picking up a part.
MUDA	Two-hand work	Work should be well defined for each of the two hands Workers have two hands that can be used in parallel in their activities. Yet, it is common that operators' work descriptions do not provide what each of her/his two hands does during the cycle time. This point is about the existence or not of such forms, not its observance.
	Walk	Walk time within a cycle time The walk time must be below 10% of the cycle time.
	Wait	Wait time within a cycle time The walk time must be below 10% of the cycle time.

FIGURE 8.3

Explanation of the eleven key workstation assessment sheet criteria.

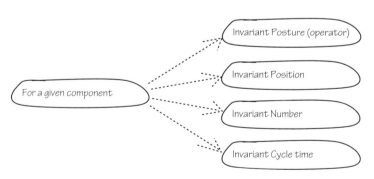

The 4 invariants (4 Is) of an efficient component supply

Each component supply solution should be designed according to the 4-Invariant rule

FIGURE 8.4

The four invariants of an efficient material supply.

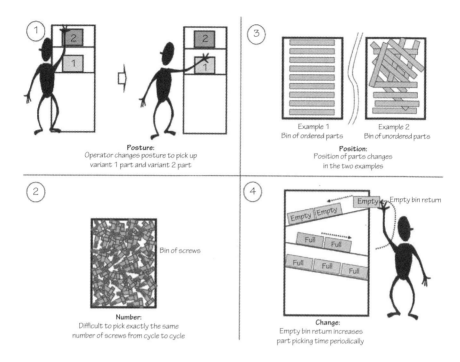

FIGURE 8.5
The four main invariants linked to part picking: constant posture, constant number, constant position, and constant time.

keep part feeding in front of the operator, then he refocused on the assessment sheet (Figure 8.2). "Well, for Mura and Muri, assessment is done for each component or tool. When the condition to be satisfied is okay, then you should fill in an O. If not, write an X. Those conditions are specified for each criterion in the second sheet you have received [Figure 8.3]. I will let you read them and get back to me if you need more explanation. The final score for each criterion here at the workstation level is the percentage of O."

Daniel stopped and breathed for a few seconds, glanced at the room to make sure that everyone was on board, then continued. "Now, when it comes to Muda, the assessment is done for the whole workstation. What are we looking for? First, what I would call the 'usual suspects' that you all know: wait and walk. There is something else pretty new that I have added, which is the use of two hands." A participant interrupted Daniel to ask: "Why is this important? And how do we achieve it?" Daniel responded, "Look, these are two excellent questions. Please allow me not to digress in order to make it easier for you to follow. I

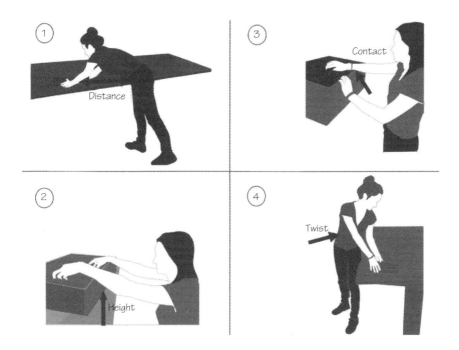

FIGURE 8.6
The four main work condition strains facing a worker when picking parts: distance to parts, height of part position, body contact, and body twist.

suggest we have a deep dive into the two-hand work issue when I finish the explanation of the forms. All right?" The participant nodded. Daniel proceeded:

Well, the conditions that each workstation must satisfy for Muda are, once again, listed in the support document [Figure 8.3]. One important point is that the final score for "two hands" is "yes" or "no." For "wait" and "walk," it is the time in seconds: recorded wait time per cycle and recorded walk time per cycle. When the form is filled out, you may want to assess another workstation belonging to the same line or the same plant. This is where you would need the second form [Figure 8.7]. Actually, it is an Excel file,* which gives a visual assessment not only of each criterion but also of the whole situation, should it be a line or a plant. There are three colors: green when it is OK (equal or better than the target), orange when it is lagging a little bit (up to 10% worse compared to target), and red when it is clearly a failure (more than 10% below the target). Also, note that each criterion aggregated at line or plant level is rated with one of the three following symbols: O

---

* File available on TheOneDayExpert.com.

Author	John B		Takt Time (sec)		WORKSTATION EVALUATION SHEET
Date	June 2012		60 sec		

	Operator Component reach Posture	Number of each Component collected by operator	Position of Components	Maximum duration periodical task	Distance to Components & Tools
How?					
Targets	100% Constant Posture	100% Constant Number	100% Constant Position	8% of Takt Time	≤ 50 cm for 100% of components
Synthesis	18% X	27% X	0% X	91% \	0% X

What? Stations	% of components OK	% of components OK	% of components OK	% of components OK	% of components OK
Workstation 1	36%	66%	38%	87%	28%
Workstation 2	100%	70%	40%	100%	90%
Workstation 3	100%	80%	60%	100%	70%
Workstation 4	98%	100%	70%	100%	70%
Workstation 5	70%	90%	67%	100%	80%
Workstation 6	95%	100%	70%	100%	80%
Workstation 7	78%	90%	11%	100%	78%
Workstation 8	90%	88%	26%	100%	74%
Workstation 9	98%	88%	70%	100%	86%
Workstation 10	40%	80%	30%	100%	80%
Workstation 11	80%	100%	70%	100%	67%

FIGURE 8.7
Synthetic Excel spreadsheet form that provides the big picture of a group of workstations belonging to the same line or to the same plant; the first raw data are from Figure 8.2. (Continued)

when it is perfectly equal or above the target; Δ when it is less than 10% below the target; X if it is worse than 10% of the target. Please note that the first row of this chart (Figure 8.7), referenced as Workstation 1, is filled out with the final score from the example on the assessment sheet just discussed [Figure 8.2]. I have added ten other workstation data as examples. If everything is clear, I now propose that we come back to the question of two-handed work.

The room acquiesced.

Well, if you remember very well, I told you at the beginning that these assessment forms were based on the most frequent shortcomings we have noticed on the shop floor. In most of the plants I have visited, folks tend to forget that they have two hands. Now, do not get me wrong. I am not taking a shot at workers; it would not be fair as there is no document that describes what the left and the right hand should be doing. Think about the potential added-value time we can harvest if workers used their two hands efficiently. Let me see; I think I have printed out an example of a form that

OVERALL ADHERENCE TO PRINCIPLES					
Target	90%		Realized	25%	

MURA				MUDA	
Height of components	Body contact while picking Components	Twist while picking Components	Two-hand Work	Walk	Wait
100% Constant Position	No body contact	No body twist	100% Defined & Applied	< 10% TT	< 10% TT
9% — X	9% — X	100% — O	0% — X	0% — X	16% — X

% of components OK	% of components OK	% of components OK	Application (YES / NO)	Walked time / Cycle	Wait time / Cycle
	63%	100%	N	10	15
95%	99%	100%	N	16	10
100%	92%	100%	N	20	5
97%	91%	100%	N	22	6
95%	98%	100%	N	32	8
67%	100%	100%	N	15	13
66%	78%	100%	N	13	10
95%	97%	100%	N	14	4
76%	90%	100%	N	16	11
90%	75%	100%	N	18	10
95%	88%	100%	N	20	13

FIGURE 8.7 (CONTINUED)

Synthetic Excel spreadsheet form that provides the big picture of a group of workstations belonging to the same line or to the same plant; the first raw data are from Figure 8.2.

shows how the operator's work can be described for both hands. Here we go. This document is called the Operator Process Chart [Figure 8.8]. The upper part shows the layout of the workstation, and right after that, below, are the work instructions for the left hand with estimated time on the left and the work instructions for the right hand with estimated time on the right. Handwritten on the bottom right is a quick estimate of potential waste. It includes any hand wait time and hold time. Why do we consider hold time as potential waste? Well, this is because it is not a value-adding task; in effect, often a simple jig or clamping system can be implemented to free the hand. Even if the operator is 100% busy (no wait time, no walk time), this Operator Process Chart shows that his or her productivity can still be improved by 45%. To be frank with you, none of our plants has such a document so far. However, we think that this is an important reservoir of productivity that we should pursue going forward.

Daniel paused a few seconds and explained that was all he had to share. There were no questions. Thomas was absent, so Daniel asked Steve if he would like to summarize the session. Steve laid down a few bullet summary points on a chart (Figure 8.9). After the summary, Daniel proposed a

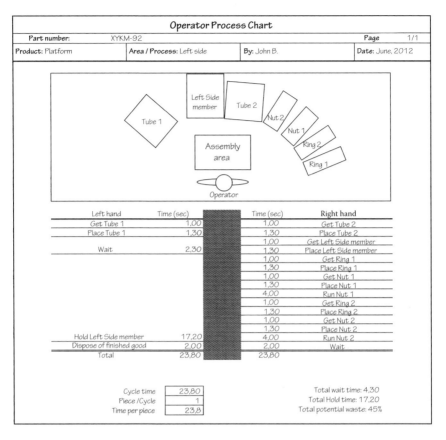

Operator Process Chart

Part number:	XYKM-92			Page	1/1
Product: Platform	Area / Process: Left side		By: John B.	Date: June, 2012	

Left hand	Time (sec)		Time (sec)	Right hand
Get Tube 1	1,00		1,00	Get Tube 2
Place Tube 1	1,30		1,30	Place Tube 2
			1,00	Get Left Side member
Wait	2,30		1,30	Place Left Side member
			1,00	Get Ring 1
			1,30	Place Ring 1
			1,00	Get Nut 1
			1,30	Place Nut 1
			4,00	Run Nut 1
			1,00	Get Ring 2
			1,30	Place Ring 2
			1,00	Get Nut 2
			1,30	Place Nut 2
Hold Left Side member	17,20		4,00	Run Nut 2
Dispose of finished good	2,00		2,00	Wait
Total	23,80		23,80	

Cycle time	23,80	
Piece /Cycle	1	
Time per piece	23,8	

Total wait time: 4,30
Total Hold time: 17,20
Total potential waste: 45%

FIGURE 8.8

Two-hand Operator Process Chart. This operator has no walk or wait time; however, there is a 45% potential for improvement by using both hands in added-value tasks.

Key Points about Workstation Assessment

- Forms are used to support Tachinbo.
- Forms are based on most common weaknesses.
- Forms are very visual (green, orange, red), and assessment is based on physically measurable data.
- Criteria belong to the three types of waste: Mura, Muri, and Muda.
- Forms lead the user step by step in a structured way to the right results.
- Use of two hands can drastically increase operator's productivity.
- Acid test for parts supply: "Close your eyes and try to pick them up."

FIGURE 8.9

Main learning points on workstation assessment sheets.

lunch break and preview of the rest of the training day. "When you return from lunch, we will move to the shop floor, where you will be able to apply today's learning on real cases. Please enjoy your food, and as we say here in France, 'Bon appétit!'"

> The acid test is that you need to be able to close your eyes and pick up exactly the part you need, when you need it. In other words, parts supply is deemed optimized when a blind person can perform the picking.

> Think about the potential added-value time we can harvest if workers use their two hands efficiently.

9

Shop Floor Application and Takeaway

Thomas returned to the room when the group was preparing for the shop floor part of the training. As expected, the meeting with the labor union had not been a cakewalk. As labor union representatives were still struggling to adapt to the so-called new doctrine, they had to digest a proposal from senior management packaged in a document called "Competitiveness Plant Power 5," which was comprised of a set of actions whose implementation was believed to move the plant to a healthy 5% operating income. This give-and-take competitiveness plan included commitment from the management side to source new products from the plant, along with investment in new machines to support a capacity increase. This was meant to ensure plant sustainability and consequently job viability for the coming years. In return, the management demanded some renouncements, including a salary freeze and an extension of the current 35-hour working week to a 39-hour working week. Under the new organization, each worker would have to work four extra hours per week with no additional compensation.

When Thomas first discussed the plan, it appeared to be an "aha moment" for a labor union whose subject of discussion thus far had merely been to negotiate the right salary raise with the management. It was no more about "How much will my raise be?" but "Will I be able to keep my job?" Years of distrustful relations with management had left the labor union mostly defiant to all the signs of the degradation of the plant situation, despite repeated warnings from the management. The denial was no longer possible, and the labor union had to face the situation. An exceptional situation requires support from the base. Thomas had therefore proposed to have every associate weigh in on his proposal. The main goal of the meeting he just concluded with the labor union was mostly to agree on the best way to organize the polling that will support or dismiss the competitiveness plan. An agreement was found. Therefore, Thomas

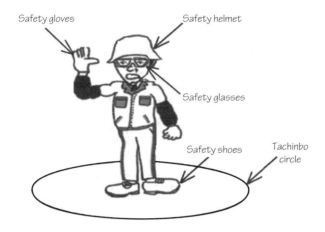

Safety gloves

Safety helmet

Safety glasses

Safety shoes

Tachinbo circle

FIGURE 9.1
Personal protection equipment (PPE) required for Tachinbo.

was in a joyful mood when he returned to the training room to carry on with the Standardized Work training.

Daniel asked the group to become prepared for the shop floor. "As we will be standing for long periods," he said, "please make sure that you're equipped with the right safety equipment; I mean safety shoes, earplugs, glasses, safety vest [Figure 9.1]. Also, remember that the shop floor is not a harmless place; therefore, be careful, look out for yourself and for your colleagues. Again, we will be standing for long hours on the shop floor. You will be focused on your observation and may also become tired. Do not loosen your attention to safety matters." Daniel also asked the plant safety manager to complete his general instructions with specific ones. He then explained that for any observation on the shop floor to be productive, everyone needed to be well equipped with required safety equipment and necessary materials (paper pad, stopwatch, camera, camcorder). Everyone needed to have a well-defined role and position as well. He concluded: "Please avoid being errant tourists on the shop floor; it is not only unproductive but also dangerous."

> Please avoid being errant tourists on the shop floor; it is not only unproductive but also dangerous.

The group decided to go back to the same assembly machines they used the day before when writing the Standardized Work forms. The plant has ten such machines. They have already written the Standardized Work

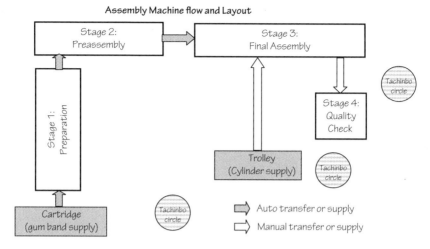

FIGURE 9.2
Layout of the machine with Tachinbo circles for observation.

forms for two of them. The group first went on the shop floor to shoot videos and pictures and to collect necessary information regarding the eight remaining machines and to write quick-and-dirty Standardized Work documents for these machines. They performed black book sharing, which led to a few modifications in the layout, and also noted some best practices to be shared between workers. They actually noticed that the machines* used the day before had the best layout. Also, they noticed that the operator was the best performer. "Did Steve choose this machine because he knew that?" Daniel asked himself. As a result, most of the changes were from the first machine to the other ones.

After the black book sharing, the group moved to the Tachinbo phase. Just like the day before, because of its size, the group was split into several teams assigned to several machines. They decided to focus on three machines to have everyone busy. Each team picked a video recorder, two people in charge of the workstation assessment forms (Figure 8.2, Figure 8.7), and two other people doing free observation to prepare for free generation of ideas. To dispatch its members, each team drew three Tachinbo circles (Figure 7.5, Figure 9.2). As underscored previously by Daniel, everyone had the requested equipment, a clear mission, and a well-defined location marked on the floor.

---

* Details are given in the previous book of this series: *Implementing Standardized Work: Writing Standardized Work Forms.*

Just-Do-It Actions to Be Implemented

- Reduce auto time of the machine by 6 seconds (this has been envisaged before but could not be performed because worker was too rushed).
- Compact the layout.
- Change job step sequence to reduce motion and walk (to be marked on the shop floor).
- Use smaller and more ergonomic equipment (height, distance, body twist) to supply cylinder. Expected savings: 12% more capacity.

FIGURE 9.3
List of just-do-it actions with a total expected gain of 12% additional capacity.

The Tachinbo lasted 2 hours. Daniel underlined that this was shortened for obvious reasons: "This is the lengthiest one we could do in this training. If you are doing Tachinbo alone, I would advise you to go as long as at least half a shift." As taught by Daniel, each team listed its proposals, made an estimate of benefits, shared and explained with videos, prioritized, then wrote the Standardized Work Improvement Story document. They used cardboard and people to perform mockups. The whole group then focused on a few just-do-it actions that they decided to perform by the end of the day (Figure 9.3). Actually, when the group looked at the Standardized Work chart, they understood quickly that something needed to be done to reduce the worker's motion. By focusing on the worker's motion, they could reduce his or her walk time, but when they looked at the Standardized Work combination chart, they noticed that the critical path was mostly constrained by the machine (Figure 9.4). This meant that any reduction of walk time (Figure 9.5) would be directly converted into wait time. Of course, this would still be a gain in terms of ergonomics but would not be visible enough to everyone in the plant, let alone managers, whether local or regional.

At this moment, John, the engineering manager, remembered that a month or so ago they noticed that they could speed up the machine and tested it but had to revert because the operator had only 3 seconds to absorb variability in the step 3 end. This did not appear to be enough (Figure 9.4). With the new job sequence, the worker would have 8 seconds at the end of step 3 and 12 seconds at the end of step 6 to synchronize with machine operation. That would be enough to absorb worker variability.

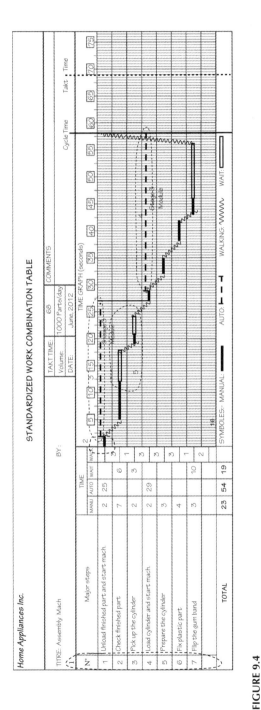

FIGURE 9.4
Before and after Standardized Work combination table: Cycle time improved by 12%. (Continued)

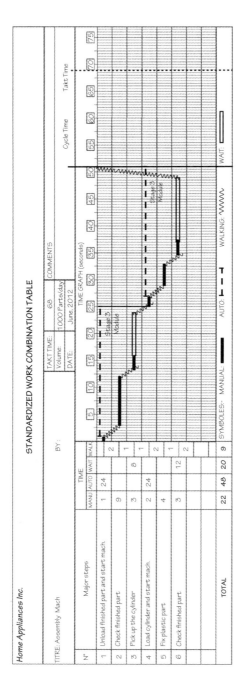

FIGURE 9.4 (CONTINUED)

Before and after Standardized Work combination table: Cycle time improved by 12%.

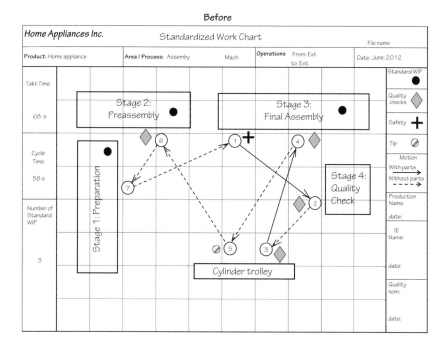

FIGURE 9.5

Before and after Standardized Work chart: more compact, less back and forth, walk reduced by 48%.

The group also noticed that for the new layout to be implemented, they needed to move from a big cylinder trolley to a smaller and more ergonomic cylinder feeder, which could be placed close to the machine, thereby preventing the operator from handling the cylinder twice. This allowed the number of job steps to be reduced from seven to six, and the operator could pick up the cylinder at the perfect height and distance and with no body twist. Very encouraged by its success, the group decided to test the new method with an operator to obtain a sense of what they called the "customer" feedback.

A workstation was quickly laid out according to the new Standardized Work chart (Figure 9.5). They also had the new setup checked and validated by the plant safety manager to make sure that they were not putting the worker at risk. Thereafter, the team leader of this assembly area, who had been part of the training from day one, was assigned to explain the new method to the worker, using Standardized Work forms. That was the best document they had available at this stage of the Standardized Work implementation.* To make it easier for a worker to adapt to the new job sequence, the group decided to mark it on the shop floor (Figures 9.6–9.8). After several runs, the worker feedback was excellent, to the point that the worker did not want the group to return to the old method. Finally, the setup that was initially designed for the test was kept after the safety manager gave it a green light.

According to the operator, the work was much easier the new way. He could feel the improvement in the work conditions. It is no surprise that he feels the change, commented Thomas: "We have halved his walking distance!" De facto after careful measurement, the group estimated that they had reduced the worker's walk distance by 48%. As Thomas commented: "This is a perfect example to start with. Everybody has won. It's a win for the worker, who will be happy with the improved work conditions. It's a win for the plant, which will enjoy increased capacity. It's a win for us because we have obtained the perfect quick win we were looking for to motivate the team!"

It was getting late, and it was time to release the group. Thomas convened all participants for a wrap-up in the training room. While walking back to the training room, Thomas reflected on the day and thought that

---

* Standardized Work forms are not destined for training. Specific documents should be used to this end. This point is addressed in the next book of the series.

FIGURE 9.6
Marking the job steps on the shop floor can ease early application of Standardized Work.

it had been a positive one overall. Referring to a quotation* that Daniel had shared with them, he mumbled: "Removing the mountain is still a long shot, but I've moved two small stones today: the successful meeting with the labor union and the positive results of today's Standardized Work training session."

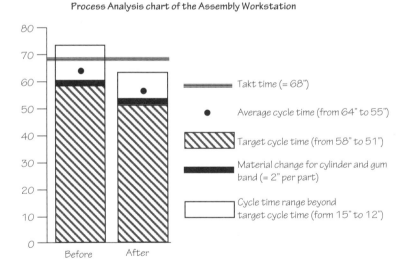

Process Analysis chart of the Assembly Workstation

Takt time (= 68")

Average cycle time (from 64" to 55")

Target cycle time (from 58" to 51")

Material change for cylinder and gum band (= 2" per part)

Cycle time range beyond target cycle time (form 15" to 12")

Before   After

FIGURE 9.7
Before and after process analysis chart: Capacity increased, and thanks to quick training and marks on the shop floor, variability was slightly reduced despite introduction of a new method.

---

* "The man who removes a mountain begins by carrying away small stones" (William Faulkner).

Main Savings Scorecard

	Saving	Comments
Ergonomics	-48% walk (2.7 km less/worker) Cylinder supply improved (height, distance, body twist)	Less walk and strain also improves worker's efficiency – worker is less tired
Safety	1 Major Key Safety Point addressed	Actions: training, communication (display)
Productivity	12% capacity increase per machine (saving: $175 k or $500 k)	Plant has 10 similar machines, 2 options: -Produce same capacity with 5 fewer operators* (saving = $175 k**) -Increase volume as part of competitiveness plan, therefore one fewer machine purchased (saving = $500k***)

*Plant needs 5 operators for each job position. For more details of the calculation, please refer to the first book of the series: Implementing Standardized Work: Measuring Operators' Performance.
**Annual cost per operator is $35 K.
*** Cost of a machine.

FIGURE 9.8
Main saving scorecard: Quick improvements help motivate the team.

Thomas thanked Daniel for his support and then continued: "I will not come back to all the steps of what you have done today. It's time to focus on the results." Thomas commented the changes on the process analysis chart (Figure 9.6) succinctly, and then displayed the saving scorecard that Steve had drafted on a flip chart (Figure 9.7). He said that the improvements were "positively surprising," then congratulated the team on their achievement and added that their "hard work has been rewarded." He also made a point of why this result was a win for everyone and concluded: "You have done a fantastic job this week, which has led to the very encouraging results in front of us. I know that you are all tired; I will let you go home now to enjoy a well-deserved rest. Oh! By the way, I have asked my assistant to print out small cards including all the steps of the process improvement for you. They are here [Figure 9.9]. Please help yourself when you leave the room." I will see you tomorrow. Drive safely!"

1-Black Books sharing

Observation:
Different operators producing the same product on the same machine are observed (see mapping in the first book).

Sharing:
The whole team shares and compares different workers' methods (use of data collection sheet, STW forms, videos).

Implementation:
A list of actions are issued and implemented right away, as most actions are just-do-its.
Result: common standard for all

2-Tachinbo

Intensive observation:
The whole team is split in several groups:
• Some groups use a workstation assessment sheet
   (take videos and pictures)
• Some groups make free observations
   (take videos and pictures)

Sharing:
Each group:

• issues a list of actions with estimated benefits
   (use mock-ups)
• shares its proposals with the whole team
   (use videos to ease understanding)

3-Prioritization

Assignment:
The whole team gives priorities to actions
(from 1 to 4 based prioritization chart –effort v. benefit).

Action plan:
The whole team agrees on an action plan. A responsible person and completion date are assigned to actions with priority below 3. Priority 4 actions are discarded.
(Use STW Improvement Story forms to communicate)

4-Execution

Focus on quick wins:
All actions classified as priority 1 and some priority 2 are implemented the same day on one machine.

Checks:
Verify that implementation produces expected results (safety, quality and cost).
For learning purposes, gaps between estimates (Tachinbo phase) actual results explained.

Communicate and celebrate:
Use Standardized Work Improvement Story to communicate
Celebrate results to motivate the squad.

5-Acting

Updating STW:
Standardized Work forms are updated to reflect improvements.

Generalization:
Improvements are extended to similar processes
(in the same plant or elsewhere).

FIGURE 9.9

Main steps of process improvement in Standardized Work implementation.

Index

About the Author

Alain Patchong is the assembly director at Faurecia Automotive Seating, France. He also holds the title of master expert in assembly processes. He was previously the industrial engineering manager for Europe, the Middle East, and Africa at Goodyear in Luxembourg. In this position, he developed training materials and led a successful initiative for the deployment of Standardized Work in several Goodyear plants. Before joining Goodyear, he worked with PSA Peugeot Citroën for 12 years, where he developed and implemented methods for manufacturing systems engineering and production line improvement. He also led Lean implementations within PSA weld factories. He teaches at École Centrale Paris and École Supérieure d'Electricité, two French engineering schools. He was a finalist of the INFORMS' (Institute for Operations Research and Management Science) Edelman Competition in 2002 and a visiting scholar at the Massachusetts Institute of Technology in 2004. He is the author of several articles published in renowned journals. His work has been used in engineering and business school courses around the world.